Design

Surface Blast Design

Calvin J. Konya

Precision Blasting Services
Montville, Ohio

Edward J. Walter

Seismological Observatory
John Carroll University

Prentice Hall, Englewood Cliffs, New Jersey 07632

Library of Congress Cataloging-in-Publication Data

Konya, Calvin J.
    Surface blast design / Calvin J. Konya, Edward J. Walter.
        p.   cm.
    Bibliography: p.
    Includes index.
    ISBN 0-13-877994-5
    1. Blasting.   2. Rock mechanics.   I. Walter, Edward J.
II. Title.
TN279.K593   1990
622'.23—dc20

Editorial/production supervision
  and interior design: *Jean Lapidus*
Cover design: *Lundgren Graphics, Ltd.*
Manufacturing buyer: *Mary Ann Gloriande*

Printed in the United States of America
10   9   8   7   6   5   4   3   2   1

ISBN   0-13-877994-5

PRENTICE-HALL INTERNATIONAL (UK) LIMITED, *London*
PRENTICE-HALL OF AUSTRALIA PTY. LIMITED, *Sydney*
PRENTICE-HALL CANADA INC., *Toronto*
PRENTICE-HALL HISPANOAMERICANA, S.A., *Mexico*
PRENTICE-HALL OF INDIA PRIVATE LIMITED, *New Delhi*
PRENTICE-HALL OF JAPAN, INC., *Tokyo*
SIMON & SCHUSTER ASIA PTE. LTD., *Singapore*
EDITORA PRENTICE-HALL DO BRASIL, LTDA., *Rio de Janeiro*

Contents

Preface

The purpose of this book is to familiarize mining and civil engineers, contractors, and blasters with the basic fundamentals of surface blast design. Blasting has advanced from an art to a science, whereby, many of the blasting variables can be calculated using simple design formulas.

This text is not meant to be a handbook or encyclopedia on blasting, rather it is meant to show a method of design which is rational and follows scientific principles. The step-by-step design methods described in this book will carry the reader from basic knowledge on explosives through considerations for proper surface blast design. The book concentrates on the fundamentals of blast design rather than details which can be learned from other texts or from field experience. Little time is spent discussing basic tie-ins of initiation systems and information of this type since it is readily available in other sources. This book will serve the beginner and the professional alike since it sorts through the vast amount of information available and puts forth a logical design procedure. The book backs up the design with some of the basic principles and theories necessary to have an understanding of why things work as they do.

The blasting industry is rapidly changing with new theories, products, and techniques. It is the goal of the authors to provide the reader with a better understanding of technology as it is today. It also points out methods of overcoming common blasting problems. This book concentrates on surface blast design. Many of the theories and procedures can be applied to underground blasting, however, it is not the intent of the authors to cover that subject in this text.

The techniques, formulas, and opinions expressed in this book are based on the experience of the authors. They should aid the reader in assessing blast designs to determine whether they are reasonable and whether they should work under average blasting conditions.

One area related to blasting which remains an art is the proper assessment of the geologic conditions at hand. Improper assessment may produce poor results in the blast. Complex geology and other factors may require changes in the design from those shown in the book, however, the methods presented would be the first step to approximate blast design dimensions which then may have to be modified to accommodate unusual local geologic conditions.

Calvin J. Konya
Edward J. Walter

Conversion Factors

Length

1 mile	=	1.6093 Km	1 Km	=	0.6214 mile
1 yd	=	0.9144 m	1 m	=	1.0936 yd
1 ft	=	0.3048 m	1 m	=	3.2808 ft
1 in	=	2.54 cm	1 cm	=	0.3937 in

Area

1 mi^2	=	2.59 Km^2	1 Km^2	=	0.3861 mi^2
1 yd^2	=	0.8361 m^2	1 m^2	=	1.196 yd^2
1 ft^2	=	0.0929 m^2	1 m^2	=	10.7639 ft^2
1 in^2	=	6.4516 cm^2	1 cm^2	=	0.155 in^2
1 acre	=	0.4047 ha	1 ha	=	2.471 acre

Volume

1 yd^3	=	0.7646 m^3	1 m^3	=	1.308 yd^3
1 ft^3	=	0.0283 m^3	1 m^3	=	35.3147 ft^3
1 in^3	=	16.3871 cm^3	1 cm^3	=	0.061 in^3

Weight

1 metric ton	=	0.9072 t	1 metric ton	=	1.1023 ton
1 lb	=	0.4536 Kg	1 Kg	=	2.2046 lb

| 1 oz | = | 28.3495 g | 1 g | = | 0.0353 oz |
| 1 grain | = | 0.0648 g | 1 g | = | 15.4324 grain |

Pressure

| 1 psi | = | 6894.75 Pa | 1 Pa | = | 145.04×10^{-6} psi |

Temperature

$$F° = 1.8 \times C° + 32 \qquad\qquad C° = (F° - 32) / 1.8$$

Surface
Blast
Design

1

Explosives Energy

INTRODUCTION

The use of explosives for blasting rock began in 1627 in the gold mines of Hungary. At that time, blastholes were hand drilled, black powder was placed into the holes, and wooden plugs were tamped tightly into the holes to confine the explosive. The use of black powder to break rock was much faster and more efficient than the traditional methods of fire setting which had been used before that time.

Blasting has become an accepted technique for rock fragmentation worldwide. Most mining and construction operations use blasting. Many types of mining ventures and construction projects would not be feasible or economically justified without the use of explosives. Many blasting procedures and techniques have been proposed since blasting began. Some techniques have passed the test of time and have been shown to work well on the average operation.

When blasting first began, there was little known about the side effects of blasting. Today, we are not only concerned with rock breakage, but also with controlling the side effects of the blast. We want to break rock safely, efficiently, and economically with the minimum amount of airblast, ground vibration, and violence.

The blasting industry has evolved from the use of a single explosive compound, black powder, to the use of many new and different explosives such as dynamites, ammonium nitrate and fuel oil, and explosive slurries.

CHEMICAL EXPLOSIVE

A chemical explosive is a compound or a mixture of compounds, which rapidly decompose, releasing large quantities of heated gases at high pressures.

High explosives go through four phases or states during a chemical reaction (Fig. 1.1). The first state is the unreacted explosive which is a solid or a solid and liquid under normal atmospheric pressures and temperatures. The second state is called the detonation state in which a high-pressure shock wave moving ahead of the reaction zone causes ionization of the explosive ingredients. The third state or explosion state is where the explosive ingredients decompose and change to gases. The resulting high pressure gases occupy the original volume of the explosive material. The fourth state of the reaction is called the expansion state. During this state, the high pressure gases that have occupied the original volume of the cartridge begin to expand, exerting forces against the rock which cause breakage to occur.

Two useful types of energy are generated by the explosive reaction. These energies will be called shock energy and gas energy.

Energy for Work

During an explosion, many types of energy are released. Useful energy capable of doing work and waste energy which can be put to no constructive use (Fig. 1.2). Two basic types of useful energy are produced when high explosives react. The first type of energy is called shock energy. The second type is called gas energy. Although both types of energy are released during the detonation process, the blaster can select explosives with different proportions of shock or gas energy to suit a particular application.

When explosives are used in an unconfined manner, such as mud capping boulders (commonly called plaster shooting) or for shearing structural members in demolition, an explosive with high-shock energy would be used. When explosives

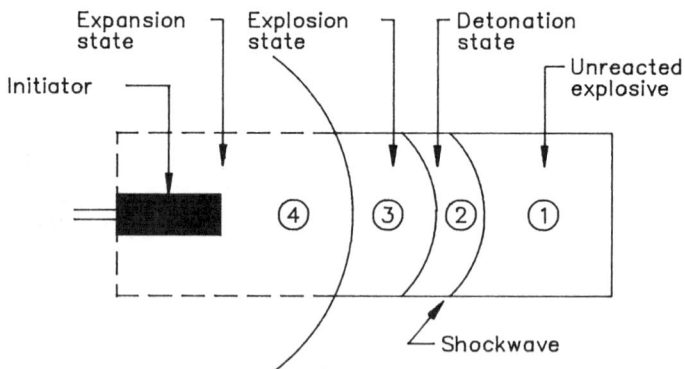

Figure 1.1   Reaction states in detonating explosives.

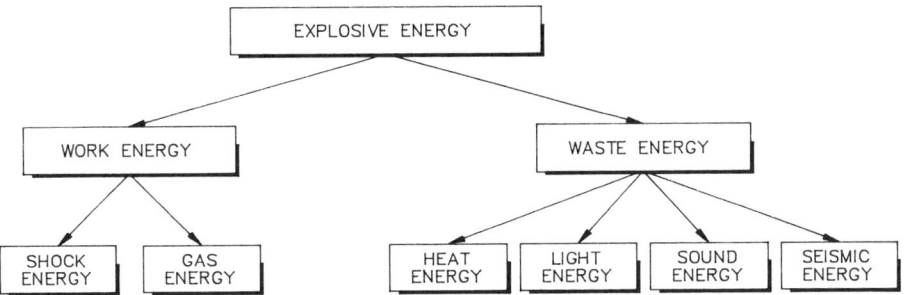

Figure 1.2 Energy release from explosives.

are used in boreholes and are confined with stemming materials, an explosive with a high-gas energy is used.

To help form a mental picture of the difference between the two types of energy, we can compare the difference in reaction of a low and high explosive. Low explosives are those which deflagrate or burn very rapidly. These explosives may have reaction velocities of two to four thousand feet per second. These explosives produce no shock energy. They produce work only from gas expansion. A typical example of a low explosive would be black powder. High explosives detonate and produce both gas pressure and shock pressure. Figure 1.3 shows a diagram of a reacting cartridge of low explosive. If the reaction is stopped when the cartridge has been partially consumed and the pressure profile examined, you can see a steady rise in pressure during the reaction until the maximum pressure is reached. Low explosives only produce gas pressure during the combustion process.

A high explosive detonates and exhibits a totally different pressure profile (Fig. 1.3). The shock pressure at the reaction front travels through the explosive before the gas energy is released. Shock energy normally produces a higher pressure than the gas pressure, however, the shock energy lasts for only a short time. After the shock energy passes, gas energy is released. The gas energy in detonating explosives is much greater than the gas energy released in low explosives. In a high explosive, there are two distinct and separate pressures. The shock pressure is a transient pressure that travels at the explosives' rate of detonation. This pressure is estimated to account for approximately 15% of the total available useful work energy in the explosion. The gas pressure accounts for approximately 85% of the useful work energy. The gas energy produces a force that is constantly maintained until the confining vessel, the borehole, ruptures.

Shock energy.   The shock energy transmitted to the rock is thought by some to result from the detonation pressure of the explosive. The detonation pressure is a function of the explosive density times the explosive reaction (detonation) velocity squared. It is a form of kinetic energy. The detonation pressure is the pressure exerted by the detonation wave propagating through the explosive column.

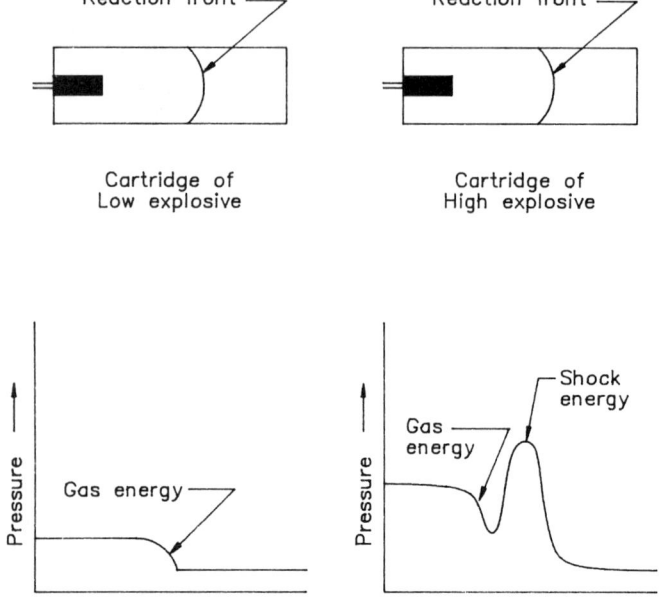

Figure 1.3 Detonation of low and high explosives.

Determination of the detonation pressure is complex. There are a number of different computer codes written to calculate this pressure. Unfortunately, the different computer codes come up with different answers. You can approximate the detonation pressure by using the equation given below.

$$P = 4.18 \times 10^{-7} \, SGe \, Ve^2/(1 + 0.8 \, SGe) \qquad (1.1)$$

where

P = the detonation pressure in kilobars
 (1 kilobar = 14,504 psi)
SGe = the specific gravity of the explosive
Ve = the detonation velocity in ft per sec

The detonation pressure is maximum in the direction of shock wave travel. The detonation pressure would be maximum in the explosive cartridge at the opposite end from where initiation occurred. The detonation pressure on the sides of the cartridge are near zero since the detonation wave does not extend to the edges of the cartridge. To get maximum detonation pressure effects from an explosive, it is necessary to place the explosives on the material to be broken and initiate it from the end opposite that in contact with the material. Laying the cartridge on its side and firing in a manner where detonation is parallel to the surface of the material to be broken reduces the effects of the detonation pressure. The material, however, is subjected to the impact caused by the radial expansion of the gases after the detonation wave has passed.

Detonation pressure can be effectively used in blasting when shooting with external charges or charges that are not in boreholes. This application can be seen in mud capping or plaster shooting of boulders (Fig. 1.4), or in the placement of external charges on structural members during demolition.

To maximize the use of detonation pressure you would want the maximum contact area between the explosive and the rock. The explosive should be initiated on the end opposite that in contact with the rock (Fig. 1.4). An explosive should be selected which has a high-detonation velocity and a high density. A combination of high density and high-detonation velocity results in a high-detonation pressure as can be calculated by using Eq. (1.1).

Gas energy.   Gas energy released during the detonation process causes the majority of rock breakage in blasting with charges confined in boreholes. The gas pressure, often called explosion pressure, is the pressure that is exerted on the borehole walls by the expanding gases after the chemical reaction has been completed. The amount of explosion pressure is related to the volume of gases liberated per unit weight of explosive and the amount of heat liberated during the reaction. The higher the reaction temperature at a constant gas volume, the higher the gas pressure. If more gas volume is liberated at the same temperature, the pressure will also increase. For a quick approximation, it is often assumed that explosion pressure is approximately one-half of the detonation pressure. Remember that this is only an approximation and conditions can exist where the explosion pressure can exceed the detonation pressure. Explosion pressures are calculated from computer codes or measured in underwater tests. Explosion pressures have been directly measured in

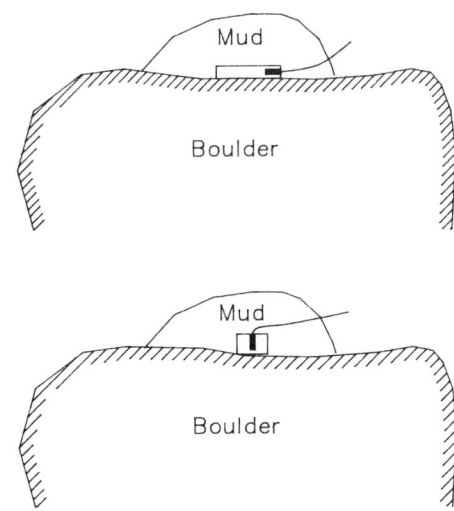

Figure 1.4 Methods of mud capping boulders.

the borehole. A review of some very basic explosives chemistry will help you to understand how explosive ingredients effect explosion pressures. The gas products formed during detonation can provide useful information on reaction, efficiency and useful energy release.

Oxygen Balance

The elements in explosives are generally considered either fuel elements or oxidizer elements (Table 1.1). Fuel elements are typically rich in carbon and hydrogen. Almost all explosives use oxygen as the oxidizer element. Nitrogen is a common component of explosives because it bonds loosely to oxygen and provides a molecular bond between the oxygen and the rest of the explosive structure. Nitrogen may represent a net gain or a net loss in energy content of the explosive, depending on its final state after reaction and upon its original bond strength in the original explosive structure.

The elements of fuel and oxidizer may be bonded into a single molecular configuration, such as in the case of TNT and nitroglycerin, which are explosive bases, or may be present in blasting agent mixtures, such as ammonium nitrate (AN) and fuel oil (FO), which are physically mixed to form the ammonium nitrate fuel oil blasting agent. The chemical bonds in the explosive must meet stringent and often conflicting requirements. They must have stability to heat, pressure and shock for safe manufacture, transportation, storage and handling. Also upon demand, the explosive must detonate promptly, completely and efficiently with maximum energy released for rock fragmentation. These results are obtained by designing chemical structures whose bonds have a high activation energy for breakage or decomposition, but are composed of elements that have in addition high-energy release upon recombination. The energy released upon initiation continues to advance the

TABLE 1.1 Explosives Ingredients

Ingredient	Chemical Formula	Function
Nitroglycerin	$C_3H_5O_9N_3$	Explosive base
Nitrocellulose	$C_6H_7O_{11}N_3$	Explosive base
Trinitrotoluene (TNT)	$C_7H_5O_6N_3$	Explosive base
Ammonium nitrate	$H_4O_3N_2$	Oxygen carrier
Sodium nitrate	$NaNO_3$	Oxygen carrier
Fuel oil	CH_2	Fuel
Wood pulp	$C_6H_{10}O_5$	Fuel
Carbon	C	Fuel
Powdered aluminum	AL	Sensitizer-fuel
Chalk	$CaCO_3$	Antacid
Zinc oxide	ZnO	Antacid
Sodium chloride	$NaCl$	Flame depressant

reaction front within the explosive and hastens the reaction, which generates large amounts of low molecular weight reaction products and additional energy. Efficient explosives produce considerable amounts of low molecular weight reaction gases rather than solids and liquids. The gases expand when heated to build pressure. Solids, on the other hand, upon heating do not build pressure. The complete net gain in energy increases product gas temperatures and these large numbers of small gaseous molecules exposed to high temperatures result in high gas pressures, which place the rock in tension and cause fracture, fragmentation, and breakage. It should be clear that a proper explosive mixture depends upon a quick release of energy and the production of low molecular weight gases. Therefore, close proximity of the fuel and oxidizer is essential. The smaller the particle size, assuming intimate mixing, the quicker the reaction will take place.

Explosives often contain ingredients other than fuels and oxidizers. For example, the explosive may have additives to decrease sensitivity or increase surface area. Chalk ($CaCO_3$) serves as an antacid as does zinc oxide. Common table salt, NaCl, is a flame depressant in permissible explosives. Coatings may exclude harmful inflow of contaminates such as water from the environment. Other additives provide anticaking properties or can be used to lower the freezing point of the mixture.

Powdered metals, such as powdered aluminum, are used to increase the heat of explosion. The aluminum itself does not contribute any low molecular weight gases upon reaction since it forms a solid Al_2O_3; however, aluminum, upon reaction, generates large quantities of heat. This heat is used to heat the low molecular weight gases from other reaction ingredients, thereby causing increased gas pressure due to the addition of temperatures to the reaction.

It is assumed that explosive reactions will go to total completion. For example, in a complete reaction of carbon and hydrogen, carbon dioxide and water are formed rather than carbon monoxide or hydroxide ions. The ideal gas law can be used as a first approximation to determine pressure resulting from an explosive reaction.

$$PV = NRT \qquad (1.2)$$

where

N = moles of gas
P = pressure
V = volume
R = the gas law constant
T = absolute temperature

Explosive reactions don't behave ideally, as would be predicted by this equation, because pressures are extremely high, therefore, correction factors must be employed to predict the behavior for the high pressure gases. Probably the simplest method of understanding and using correction factors would be with the equation:

$$PV = ZNRT \tag{1.3}$$

where

$$Z = \text{compressibility factor}$$

The compressibility factor Z is obtained from charts which require the knowledge of reduced temperatures and reduced pressures resulting from the gas.

When oxygen balanced explosives detonate, all fuel elements oxidize in the reaction, and those reactions must go to completion. There should not be excess oxygen available to oxidize the nitrogen gas liberated from the reaction. If the nitrogen gas is oxidized, a rust-colored fume is present which actually reduces the total energy and the total available work of that explosive.

Problems from poor mixtures.   Explosives must undergo a change so that the energy can be released and used. The chemical energy stored in explosives must be contrasted with the physical energy released when water falls or when steel springs uncoil. The water and steel spring are not changed chemically, but have a different location in a gravitational force field or a different state of compression. The amount of chemical energy stored in a chemical explosive is related to the nature and balance of the constituents. In practice, the chemical reaction, which released energy, requires that a fuel or reducing agent must be present along with an oxidizing agent. In an ideal reaction, hydrogen is completely reacted to form H_2O or water, the carbon completely reacts to CO_2, and nitrogen becomes N_2, nitrogen gas. The other materials react to form the thermodynamic equilibrium reaction products.

Most explosives are formulated to balance the fuel and oxygen to avoid a deficiency or an excess of oxygen. A deficiency of oxygen would lead to low energy release, slow energy release, a difference in the number of small molecules, lower temperature, and thereby lower pressures. This could also result in the formation of poisonous products such as carbon monoxide (CO). An excess of oxygen would lead to extreme sensitivity, lower energy release, and to the formation of wasteful or poisonous oxides of nitrogen: N_2O_5, N_2O_3, NO_2, N_2O, NO. The concept of fuel and oxygen balance can best be illustrated by Ex. 1.1.

Example 1.1:

Consider a mixture of ammonium nitrate and fuel oil. Ammonium nitrate (AN) has the chemical formula: NH_4NO_3, which shows the constituents and their bonding.

Rewriting the formula gives $N_2H_4O_3$ and in table form as follows:

Elements	Reaction Product Desired	Oxygen Needed
N_2	N_2	None
H_4	$2H_2O$	2
O_3	None	None

Deficiency or excess: $3 - 2 = 1$ excess of oxygen.

Fuel oil (FO) is added to an AN to provide necessary fuel. This fuel balances the excess oxygen in unmodified AN. For example, FO may be considered for our purposes to have the formula, CH_2. Thus, if the CH_2 is written in table form and an oxidation reaction occurs:

Elements	Reaction Product Desired	Oxygen Needed
C	CO_2	2
H_2	H_2O	1

Zero (0) = -3 (oxygen deficiency).

The result is a deficiency of three oxygens per CH_2 unit. Since each AN has a one oxygen excess, three AN units are required to balance one FO unit in ANFO mixtures:

$$3 \text{ AN} + 1 \text{ FO} \rightarrow CO_2 + 7H_2O + 3N_2$$

From this balanced reaction, one can see that it takes three molecular weights of ammonium nitrate and one molecular weight of fuel oil to produce the ideal gases previously discussed.

Energy release calculations.   To estimate total available energy released during an explosive reaction, the reaction is assumed to lose or gain no energy from the outside, in that all the heat energy released is assumed to heat the explosive products. The reaction is assumed to go to reaction completion or to be ideal. Because the pressures developed are a direct function of the number of molecules and the temperatures of the gases, the explosive work potential is directly related to the amount of heat released (Q_e).

Energy released can be readily calculated. The heat released is the difference between the total heat of formation for the reaction products (Q_p) and the total heat of formation for the reactants (Q_r). Thus $Q_e = Q_p - Q_r$. The respective heats of formation of various compounds are given in Table 1.2. Additional values can be found in the Handbook of Chemistry and Physics (1962).

For making calculations using heats of formation, one must recognize the assumption that no heat is required to form any pure elements. Therefore, C, N, H, O, Al, and so on, have a zero heat of formation. If heat is released during reaction, the compound is said to have a negative heat of formation (exothermic). If heat must be added to the reaction to cause it to occur, the compound is said to have a positive heat of formation (endothermic). All explosive reactions must be exothermic. A temperature of $298°$ K and a pressure of 760 mm are considered when listing heats of formation in data tables. The procedure for calculating the value for an explosive can be illustrated by using the example mixture of ammonium nitrate and fuel oil. Example 1.2 illustrates this procedure substituting the proper numbers. First the balance reaction is written. The heats of formation for the various products are determined. The heats of formation of the products and reactants are determined from Table 1.2. Subtract the product heat from the reactant heat and the result is the heat of explosion. The heat of explosion must be divided by the number of grams in the mixture in order to normalize the reaction to a one gram or unit weight basis.

TABLE 1.2   Heats of Formation for Selected Chemical Compounds

Compound	Formula	Mol:Wgt.	O_p or Q_R (Kcal/Mole)
Corundun	Al_2O_3	102.0	−399.1
Fuel oil	CH_2	14.0	− 7.0
Nitromethane	CH_3O_2N	61.0	− 21.3
Nitroglycerin	$C_3H_5O_9N_3$	227.1	− 82.7
PETN	$C_5H_8O_{12}N_4$	316.1	−123.0
TNT	$C_7H_5O_6N_3$	227.1	− 13.0
Carbon monoxide	CO	28.0	− 26.4
Carbon dioxide	CO_2	44.0	− 94.1
Water	H_2O	18.0	− 57.8
Ammonium nitrate	$N_2H_4O_3$	80.1	− 87.3
Aluminum	AL	27.0	0.0
Carbon	C	12.0	0.0
Nitrogen	N	14.0	0.0
Nitrogen oxide	NO	30.0	+ 21.6
Nitrogen dioxide	NO_2	46.0	+ 8.1

Since, on the other hand, it is customary to use kilograms as the unit of measure, the heat of formation is multiplied by 1,000 grams per kilogram to get units in kilocalories per kilogram.

Example 1.2 Heat of Explosion Calculations for Ammonium Nitrate and Fuel Oil

(From Example 1.1)

$$3AN + 1FO \rightarrow CO_2 + 7H_2O + 3N_2$$

$$3AN + 1FO = \text{Reactants}$$
$$3(-87.3 \text{ Kcal}) + (-7 \text{ Kcal}) = \text{Heat of formation of reactants } (Q_r)$$
$$-268.9 \text{ Kcal} = Q_r$$

$$CO_2 + 7H_2O + 3N_2 = \text{Products}$$
$$(-94.1 \text{ Kcal}) + 7(-57.8 \text{ Kcal}) + 3(0 \text{ Kcal}) = \text{Heat of formation of product } (Q_p)$$
$$-498.7 \text{ Kcal} = Q_p$$

$$Q_p - Q_r = Q_e = \text{Heat of explosion}$$
$$-498.7 \text{ Kcal} - (-268.9 \text{ Kcal}) = -229.8 \text{ Kcal} = Q_e$$

Molecular weight of compound from Table 1.2

$$3(AN) + 1FO = 3(80.1g) + 1(14g) = 254.3g$$

To calculate Kilocalories per Kilogram (Kcal/Kg)

$$(-229.8 \text{ Kcal}/254.3g) \times (1000g/\text{Kg}) = -903.7 \text{ Kcal/Kg}$$

Heat of explosion $= -903.7$ Kcal/Kg

Example 1.3 considers the addition of a small amount of aluminum to an ammonium nitrate/fuel oil mixture. The same steps as indicated earlier are followed to get a balanced chemical reaction and the heat of explosion. Powdered metals produce a high value for heat release and form solid oxides rather than gaseous products. The resulting gas pressure cannot be linearly related to heat release.

Example 1.4 illustrates the reaction of ammonium nitrate and insufficient quantities of fuel oil, which would produce a positive oxygen balance.

If you want to calculate the weight proportion of each ingredient needed to make the balanced mixture, you need only take the molecular weight of any ingredient and divide by the total molecular weight of the compound. That ratio is the amount of the ingredient in the total weight of mixed compound. See Ex. 1.5 for a worked solution.

Example 1.3 Heat of Explosion Calculations for Aluminized ANFO

Compound	Formula	Reaction Products Desired	Oxygen Surplus/Deficiency
AN	$N_2H_4O_3$	N_2, $2H_2O$	$+1$
FO	CH_2	CO_2, H_2O	-3
AL	AL	$0.5AL_2O_3$	-1.5

$$AN + FO + AL \rightarrow CO_2 + H_2O + N_2 + AL_2O_3$$

Need 4.5 AN for FO + AL
Balanced reaction

$$4.5\ N_2H_4O_3 + CH_2 + AL \rightarrow 4.5N_2 + 10H_2O + CO_2 + 0.5AL_2O_3$$
4.5AN + FO + AL = Reactants
$4.5(-87.3\ Kcal) + (-7\ Kcal) + (0\ Kcal)$ = Heat of formation of reactants (Q_r)
$-399.85\ Kcal = Q_r$
$4.5N_2 + 10H_2O + CO_2 + 0.5AL_2O_3$ = Products
$4.5(0\ Kcal) + 10(-57.8\ Kcal) + (-94.1\ Kcal) + 0.5(-399.1\ Kcal)$ = Heat of formation of products (Q_p)

$$871.65\ Kcal = Q_p$$
$$Q_p - Q_r = Q_e = \text{Heat of explosion}$$
$$-871.65\ Kcal - (-399.85\ Kcal) = -471.8\ Kcal = Q_e$$
Molecular weight of compound from Table 1.2

$$4.5(AN) + (CH_2) + (AL) = 4.5(80.1g) + 14g + 27g = 401.45g$$

To calculate Kilocalories per Kilogram (Kcal/Kg)

$$(-471.8\ Kcal/401.45g) \times (1000g/Kg) = -1175.24\ Kcal/Kg$$

Heat of explosion = -1175.24 Kcal/Kg

Example 1.4 Heat of Explosion Calculations for Underfueled ANFO

$$6AN + FO \rightarrow CO_2 + H_2O + N_2 + NO + NO_2$$
$$6N_2H_4O_3 + CH_2 \rightarrow CO_2 + 13H_2O + 5N_2 + NO + NO_2$$
$$6AN + FO = \text{Reactants}$$

$6(-87.3 \text{ Kcal}) + (-7 \text{ Kcal}) = \text{Heat of formation of reactants } (Q_r)$
$-530.8 \text{ Kcal} = Q_r$
$$CO_2 + H_2O + N_2 + NO + NO_2 = \text{Products}$$
$(-94.1 \text{ Kcal}) + 13(-57.8 \text{ Kcal}) + 5(0 \text{ Kcal}) + 21.6 \text{ Kcal} + (8.1 \text{ Kcal}) = \text{Heat of forma-}$
tion of products Q_p
$$-815.8 \text{ Kcal} = Q_p$$
$$Q_p - Q_r = Q_e$$
$$-815.8 \text{ Kcal} - (-530.8 \text{Kcal}) = -285 \text{ Kcal} = Q_e$$

Molecular weight of compound

$$6(80.1g) + 14g = 494.6g$$

To calculate heat of explosion in Kilocalories per Kilogram (Kcal/Kg)

$(-285 \text{ Kcal}/494.6 \text{ g}) \times (1000 \text{ g/Kg}) = -576.2 \text{ Kcal/Kg}$

Heat of explosion $= -576.2$ Kcal/Kg

Example 1.5 Determination of Pounds of Ingredients and Products for Oxygen Balanced Mixture of 100 LB of ANFO

(From Ex. 1.1)

$$3AN + 1FO = CO_2 + 7H_2O + 3N_2$$

Molecular weight of compound (MWC)
$3AN + FO = 3(80.1g) + 1(14g) = 254.3g$
For 100 lb mixture

Pounds of AN $= (3AN/MWC) \times 100 = (3(80.1g)/254.3g) \times 100 = 94.5$ lb

Pounds of FO $= (1(FO/MWC) \times 100 = (14g/254.3g) \times 100 = 5.5$ lb

Pounds of $CO_2 = (CO_2/MWC) \times 100 = (44g/254.3g) \times 100 = 17.3$ lb

Pounds of $H_2O = (7(H_2O)/MWC) \times 100 = (7(18g)/254.3g) \times 100 = 49.6$ lb

Pounds of $N_2 = (3(N_2)/MWC) \times 100 = 3(28g)/254.3g) \times 100 = 33.1$ lb

Identification of Problem Mixtures

There are visual indicators of proper and improper energy release. The gases released are indicators of reaction efficiency and associated energy release. When light gray colored steam is released, oxygen balance is near ideal and maximum energy is released. When gases are either yellow or rust colored, they indicate an inefficient reaction, one that may be due to an oxygen positive mixture. Oxygen negative mixtures produce dark gray gases and often leave carbon on borehole walls.

The visual indicators of improper mixture can be put into systematic classification for easy field identification. Figure 1.5 is an easy to use guide which relates the elements and the reaction products. You will notice that below the reaction products there are notes in parentheses which indicate the color of each product. The elements can react to form one of three types of reactions. All commercial explosives should be nearly oxygen balanced. Therefore, upon detonation ideal reaction products should form only light gray fumes. Unfortunately, in the field it is not uncommon to see other colors of fumes after reactions have occurred.

Whenever colored fumes are seen after a blast, it is a sign that the explosive reaction was inefficient and some useful energy was lost. The problem should be corrected immediately since it can lead to dangerous blasting conditions. When insufficient energy is released, the blastholes may not have sufficient energy to break the rock properly in front of the hole. The blastholes are, therefore, underloaded for the amount of rock present. Underloaded blastholes can lead to violence, such as airblast, and flyrock. Underloaded holes also cause higher than normal ground vibration, unwanted breakage behind the blastholes and blastholes may not break to the desired depth.

Visual observation of colored gases after a blast indicate that problems exists, however, the magnitude of the problems cannot be judged by the intensity of the color alone.

In order to get a better understanding of the importance of the proper fuel oxidizer ratio in field blasting, let us look at the amount of energy loss that results

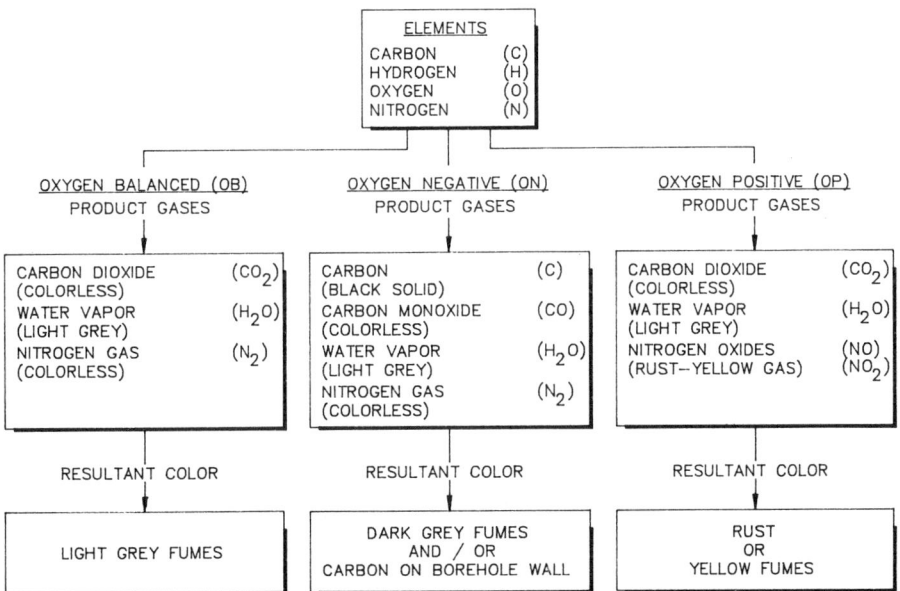

Figure 1.5 Identification of problem mixtures.

from a simple yet extensively used explosive. Figure 1.6 relates the energy loss of
ammonium nitrate and fuel oil to the percent oil which was placed in the mixture.
Theoretically, to get the optimum balance, there should be 5.7% fuel oil and 94.3%
ammonium nitrate. Manufacturers will normally try to get approximately 6% fuel
oil in the mixture. You will notice in Fig. 1.6 that as the oil content decreases from
the desired 5.7% there is a rapid energy loss. If only 1% fuel oil is present, the heat
release from the explosion drops by approximately 40%. Excess oil does not produce the same magnitude of losses. If the equipment which mixes the fuel oil and
the ammonium nitrate is not precise, you would always be better served by having
excess oil rather than insufficient oil. The strength of the explosive and the energy
delivered in the field is not a linear relationship with heat released, therefore, using
the heat released chart alone as an indicator of strength or energy loss is misleading.
From a field performance standpoint, judging efficiency by heat release alone for an
ammonium nitrate fuel oil mixture with a low oil percentage over estimates the
useful energy release.

Waste Energy

Explosives release energy and produce rock fracturing, plastic deformation,
and elastic deformation of the rock.

The blasting energy that causes elastic deformation produces the stress waves
(body waves) which move through the rock mass and can cause further fracture
upon rebound from discontinuities. The elastic deformation also causes the seismic
waves that are troublesome since at high levels they can be damaging to structures
and annoying to people.

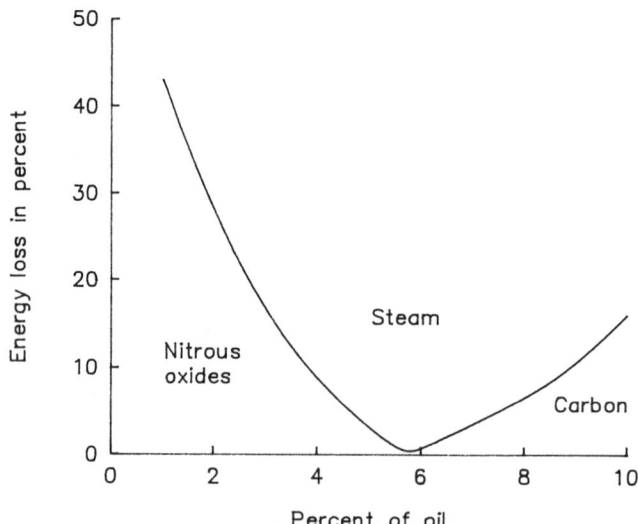

Figure 1.6   Energy loss in ANFO.

The shock energy and gas energy, which we have previously discussed, can cause rock fracture when properly applied. The explosive reaction also produces energy which does not in itself, lead to fracture and does no useful work in the blasting process. This energy is commonly called waste energy and it consists of light, heat, sound, and seismic energy.

The seismic and sound energy are those waste energy sources that commonly contribute to the problems associated with blasting.

Seismic energy. Seismic energy produces waves that are the transmission of energy through the solid earth. Other types of energy transmission by waves are sound waves, light waves, and radio waves. Earthquakes are a natural source of seismic waves, and the science that studies earthquakes is *Seismology*, the name being derived from the Greek word *seismos* meaning to shake. Man's activity has created many artificial sources of seismic waves and when these man-made seismic waves can be felt, they are referred to as "vibration."

A "vibration problem" has developed as the use of explosives has increased. What this means is that blasting produces seismic waves of sufficient motion that people can feel them. Blasting is not the only source of vibration. Many industrial activities such as forging operations, heavy press operations and construction activities such as pile driving, and jack hammering are a few that generate vibration. People are disturbed, concerned, even fearful when they feel the vibration. A confrontation results that is known as the "vibration problem."

Serious investigation of the vibration problem began in the early thirties. This became the basis for continuing research by many scientists and important results have been forthcoming. The vibration problem remains the subject of continuing research. This has produced a vast bank of information and a high level of technology. A brief discussion of seismic waves is a good point to begin an examination of seismic energy.

There are two classes of seismic waves: body waves and surface waves.

Body waves. These waves are called body waves because they travel into and through the rock mass, penetrating down into its interior. Body waves are of two types: compressional waves and shear waves. A compressional wave produces alternating compression and dilatation in the direction of wave travel. This is a push-pull kind of motion similar to what occurs in a stretched spring. In seismology, compressional waves are call P-waves. Solids, liquids, and gases transmit compressional waves. The shear wave is a transverse wave that vibrates at right angles to the direction of wave travel. This transverse type motion can be demonstrated by flexing a rope. The rope moves up and down, but the wave travels outward toward the other end. In seismology, shear waves are called S-waves. Liquids and gases do not transmit shear waves. They are transmitted only by solids. Compressional waves and shear waves travel at different speeds with the compressional wave traveling faster.

Surface waves.   Surface waves travel along the outer surface layer of rock and do not penetrate into the rock mass. The wave motion decreases with depth so that at a depth of approximately one wave length, the motion has diminished to zero. Physical and geometrical conditions prevent them from traveling into the interior of the rock mass. Surface waves are larger than body waves but travel slower. Significantly for the vibration problem, surface waves are the large energy carriers and produce the largest motions. The motion for compressional waves and shear waves is shown in Figs. 1.7 and 1.8.

Formation of Seismic Waves

Seismic waves exist because of the elastic nature of rock material. *Elasticity* is a property of matter by which a material returns to its original shape or size if it is deformed. A stretched rubber band demonstrates elasticity when it springs back to its original length as it is released. Rock materials are highly elastic and produce strong elastic or seismic waves. When deformation of a solid occurs, it shows up as a change in volume which is a compression, or a change in shape which is a shear.

Materials do not deform easily, they resist. Try compressing a rock. The resistance to compression is measured by the modulus of incompressibility, or the bulk modulus. The resistance to change in shape is measured by the modulus of rigidity, or the shear modulus. These elastic moduli are important in the development of the seismic wave equations, which led to calling the waves compressional waves and shear waves.

Blasting, by its very nature, will always produce vibration or seismic waves. This becomes obvious when the purpose of blasting is examined. The purpose of blasting is to fracture rock. Thus, an amount of energy must be supplied that will exceed the strength of the rock, that is exceed the elastic limit. The rock will then fracture and continue fracturing until the energy is used up and falls to a level less than the strength of the rock. Fracturing then stops. The remaining energy that is below the elastic limit can only deform the rock. This deformation is transmitted from rock particle to rock particle. This is what constitutes a seismic wave. A schematic representation of compression and shear is shown in Figs. 1.9 and 1.10.

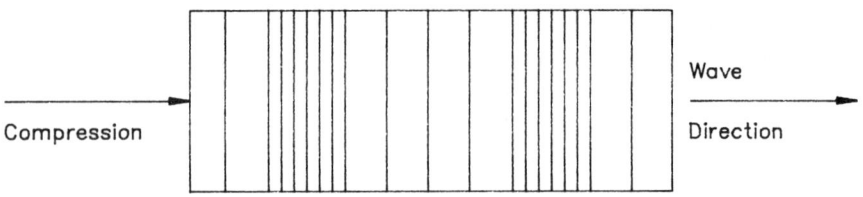

Figure 1.7   Compressional wave.

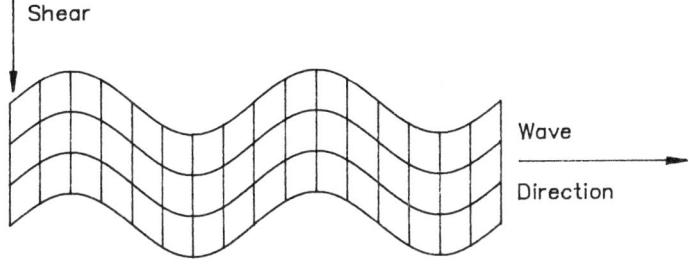

Figure 1.8   Shear wave.

Sound.   Blasting generates sound waves that are heard as loud noise. It also generates subaudible noise that cannot be heard. Sound is the transmission of energy through the atmosphere. If there were no atmosphere, there would be no sound. Sound is not transmitted in a vacuum, it requires a medium of transmission.

Blasting sound represents waste energy in the sense that similar to seismic energy, it does not fracture rock which is the primary function of blasting.

What kind of waves are these sound waves? From a physical point of view, the atmosphere is a liquid which resists change in volume but not change in shape. It has volume elasticity but not shear elasticity. Liquids, therefore air, transmit compressional waves but not shear waves. Sound waves are compressional waves.

The velocity of sound is a function of temperature. If the air temperature increases, sound velocity increases. If the air temperature decreases, sound velocity decreases. This has significant bearing on the transmission of sound through the atmosphere sometimes causing the direction of the sound to change and energy concentrations to occur. At normal conditions, the velocity of sound is approximately 1100 ft/sec.

This waste energy called noise is generated in a number of ways when a blast is fired. Noise arises when the rock fractures and the gas pressure in the borehole is vented to the atmosphere, when stemming blowout occurs, when the rock face is displaced, and when displacement occurs around the boreholes. Any or all of these can occur on a given blast.

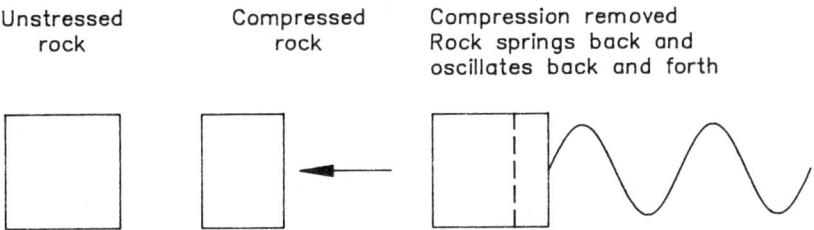

Figure 1.9   Compressional deformation and compressional wave.

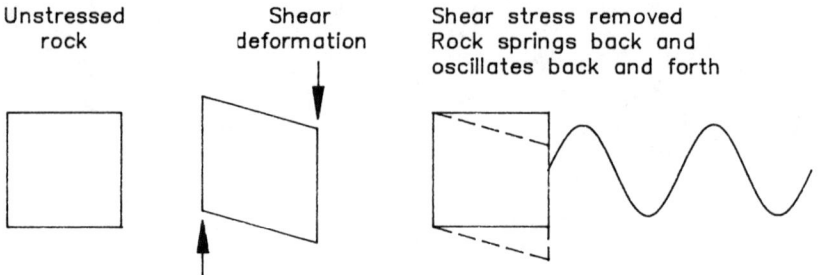

Unstressed rock   Shear deformation   Shear stress removed Rock springs back and oscillates back and forth

Figure 1.10   Shear deformation and shear wave.

PROBLEMS

1. Determine the detonation pressure for ammonium nitrate ($SGe = 0.8$) found in a 3-in. diameter pipe with a detonation velocity of 12,000 feet per second (fps).
2. Determine the detonation pressure of PETN ($SGe = 1.6$) with a detonation velocity of 21,000 fps.
3. Determine the chemical equation for an oxygen-balanced mixture of ammonium nitrate and pure carbon.
4. Find the heat release for the mixture in Prob. 3.
5. Determine the chemical equation for an oxygen-balanced mixture of ammonium nitrate and TNT.
6. Find the heat release for the mixture in Prob. 5.
7. Find the number of pounds of ingredients and products for 100 lb of mixture in Prob. 3.
8. Name the types of seismic waves and their subdivisions.
9. Seismic waves are produced because of a property of matter which is called _____.
10. The resistance of rock to deformation is called _____.
    (a) The resistance to compression is the _____ _____.
    (b) The resistance to change in shape is the _____.
11. Why does blasting cause seismic waves?
12. Why is the seismic wave energy called waste energy?
13. What kind of waves are sound waves?

REFERENCES

ASH, R.L., "The Mechanics of Rock Breakage," Parts I, II, III, and IV. *Pit and Quarry*, 56, no. 2 (Aug. 1963), pp. 98–112; no. 3, (Sept. 1963, pp. 118–123; No. 4, Oct. 1963, pp. 126–131; No. 5, Nov. 1963), pp. 109–111, 114–118.

COOK, M.A., "Explosive—A Survey of Technical Advances," *Ind. and Eng. Chem.*, 60, no. 7 (July 1968), pp. 45–55.

Cook, M.A., "The Science of Industrial Explosives," Ireco Chemicals, Salt Lake City, UT, 1974, pp. 449.

Hodgman, C.D., ed., *The Handbook of Chemistry and Physics,* 43rd ed., Chemical Rubber Publishing Co., Cleveland, 1962.

Leet, L.D., *Earth Waves,* 7th ed., Harvard University, Cambridge, MA, 1950.

Macelwane, J.B., *Theoretical Seismology,* John Wiley & Sons, New York, 1936.

Richter, C.F., *Elementary Seismology,* San Francisco, CA: W.H. Freeman and Co., 1958.

2

Breakage Mechanisms

ROCK FRAGMENTATION THEORIES

Many theories have been proposed as to how and why rock breaks as a result of an explosion. Two separate and distinct mechanisms occur. The explosion first induces a *stress wave*, which is a compressional wave, into the rock surrounding the charge. This stress wave moves through the material at near the speed of sound or compressional wave velocity in the rock. After the wave passes through the rock, the gas pressure in the blasthole again stresses the rock. The different breakage theories placed different weights on the role of the stress wave (shock energy) and the gas pressurization. Some theories conclude that the stress waves, shock energy, is responsible for the majority of breakage, while the gas pressurization effect on breakage is considered negligible. Other theories consider the gas pressurization to be dominate in the breakage process with the stress waves causing negligible breakage. It is not the intent of the author to describe all past theories of breakage, however, a few observations as to the general weaknesses of some past theories will be made.

Part of the confusion surrounding the understanding of the breakage process is because there are two types of energy produced. The different types of energy cause different effects depending on how the explosive is applied, confined in a borehole or unconfined on the rock surface. Another factor contributing to the confusion is that much of the research has been done in small scale models in the laboratory, and the magnitude and ratio of the shock and gas energy are not consistent with explosive application in the field.

There is no doubt that in the laboratory with small models, fragmentation can be achieved with either shock energy or gas energy. Field results, however, demonstrate that to cause extensive rock fragmentation from shock energy alone would require many times the explosive loads normally used in the field. Full scale field research has conclusively shown that some laboratory research was flawed by experimental procedures and magnitudes of explosives used. Laboratory research results are often misinterpreted because true field conditions cannot be accurately modeled—especially shock and gas energy ratios and charge geometry.

It should be obvious that when charges are placed on or near the surface of models and the explosive is relatively unconfined, little work can result from the gas pressure. Any useful work accomplished would, by necessity, be from the shock energy induced into the rock. Conversely, when charges are placed in a borehole, the gas pressure does dominate and the effect of the stress wave is minimal. The reason for this apparent reversal is because approximately 85% of the useful work energy released during the detonation is from the gas pressure, while only approximately 15% of the energy goes into the stress waves. If this energy partitioning is kept in mind during the following discussion, it will be easier to understand the breakage process.

Konya's Breakage Theory

When charges are detonated under field conditions in blastholes, the following stages of breakage result:

STAGES

1. After detonation, the stress wave conditions the rock by causing microfractures to result at the borehole walls and at some discontinuities between the hole and the face (Fig. 2.1). The stresses produced are of insufficient magnitude to cause spalling at the face under normal field conditions. The effect of the stress wave is minimal when using charges confined in boreholes. Less than 15% of the breakage results from shock energy and the resulting stress wave.

2. After the stress wave has passed, the expanding gases cause pressurization of the blasthole which produces radial cracks on the borehole walls. The radial cracks form as a result of the hoop stress generated by the gas expansion. Gas pressure drives the radial cracks through the rock to the face. The direction of the radial crack growth is controlled by the amount of resistance in front of the hole. Normal resistance causes radial cracks to concentrate toward the face. Excessive resistance causes somewhat symmetrical cracking around the hole,

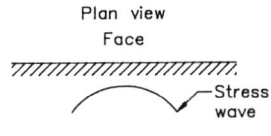

Figure 2.1 Breakage process, Stage 1.

Plan view
Face

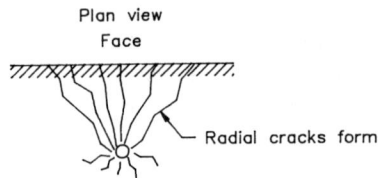

— Radial cracks form

Figure 2.2 Breakage process, Stage 2.

both in front and behind the blasthole. The majority of the fractures form parallel to the axis of the boreholes during this stage of breakage (Fig. 2.2).

3. After the radial cracks form, the high-pressure gases penetrate the radial crack network. The radial cracks are pressurized approximately 60% of their length from the hole to the face before any face movement occurs (Fig. 2.3).

4. Face movement begins and flexural failure occurs as a result of the bending of the rock mass in two planes. Both are in the plane of the charge diameter and in the plane of the charge length (Fig. 2.4).

   The bending action in the plane of the charge length is similar to bending a beam. High faces and long boreholes break easier than blasting under similar conditions on low faces which are stiffer and more difficult to break. Both the charge geometry and the explosive amount are critical in the breakage process. The face displacement at any stress level is controlled by the modulus of elasticity of the rock (Young's modulus) (Fig. 2.5).

There are many factors that contribute to the breakage process such as mid-air collisions of broken rock. We will, however, emphasize only those mechanisms that play a dominate role in the breakage process. Those mechanisms are the ones we want to control in order to produce the desired end results.

We can now discuss how rock breaks in field applications with both unconfined and confined charges.

Unconfined charges.   Unconfined charges such as those placed on boulders and subsequently detonated produce shock energy which is transmitted into the boulder only at the point of contact between the charge and the boulder (Fig. 2.6). Since most of the charge is not in contact with the boulder, the majority of the useful explosive energy travels radially outward into space and is wasted. This wasted energy manifests itself in excessive airblast. Gas pressure can never build since the

Plan view
Face

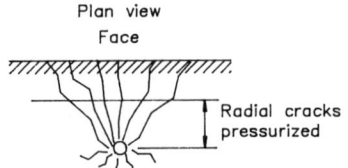

Radial cracks
pressurized

Figure 2.3 Breakage process, Stage 3.

Plan view   Section view

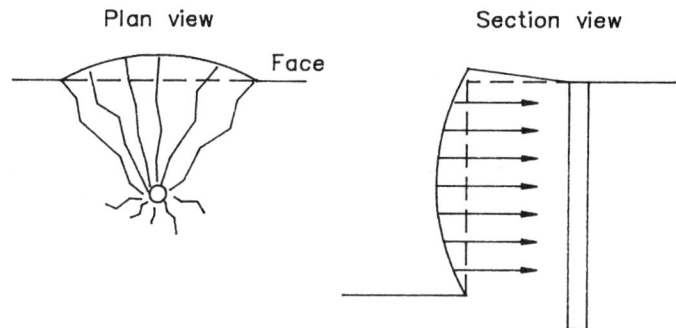

Figure 2.4   Breakage process, Stage 4.

charge is totally unconfined, therefore, gas energy does little work. Only a small amount of the useful potential energy of the explosive is utilized for rock breakage when high explosive charges are placed unconfined on boulders.

Let us compare two methods of boulder blasting, one in which the explosive charge is placed in a drill hole, and confined with clay to the hole collar, and in the second case a charge placed unconfined on top of the boulder. You would find that it requires many times the amount of explosive on top of the boulder to obtain the same fragmentation as the confined charge within the borehole. The surface charge breaks using only the shock component of total energy while the smaller confined charge utilizes primarily the gas energy component.

Figure 2.5   Bending action at face. (Courtesy of Nitro Nobel AB, Gyttorp, Sweden)

Figure 2.6 Unconfined charge on boulder.

Plaster shooting.   *Mud capping* (plaster shooting) is a commonly used method to increase breakage with surface charges. The method utilizes a thin layer of mud which is placed on the boulder with the explosive cartridges pressed firmly into the mud. The charge is covered by mud preferably shaped into a hemispheric mound. The technique causes the explosive charge to exert more downward force into the boulder than if the mud was not used. You could conclude that the gas confinement offered by a few handfuls of mud helped in the breakage process. Common sense would indicate, however, that this would not be logical since a few handfuls of mud could not significantly resist pressures near one million psi. The mud actually forms a wave shaper, whereby some of the wasted shock energy, which would normally go off into space is reflected back into the boulder (Fig. 2.7). Wave shapers have been known and used for decades in explosive metal forming operations.

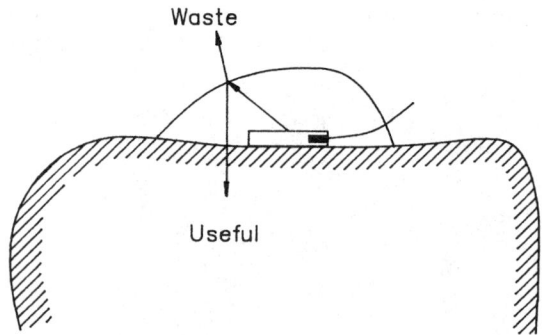

Figure 2.7 Reflected and waste energy in mud cap blasting.

Confined Charges in Boreholes

Three basic mechanisms contribute to rock breakage with charges confined in boreholes. The first and least significant mechanism of breakage is caused by the shock wave. At most, the shock wave causes microfractures to form on the borehole walls and initiates microfractures at discontinuities in the burden. This transient pressure pulse quickly diminishes with distance from the borehole and outruns the fracture propagation since the propagation velocity of the pulse is approximately 2.5 to 5 times the maximum crack propagation velocity.

The two major mechanisms of rock breakage results from the sustained gas pressure in the borehole. When the solid explosive is transformed into a gas during the detonation process, the borehole acts similar to a cylindrical pressure vessel. Failures in pressure vessels, such as water pipes or hydraulic lines, are similar to this mechanism of rock breakage. When a pressure vessel is overpressurized, the pressure exerted perpendicular to the confining vessel's walls will cause a fracture to occur at the weakest point in the pressure vessel. In the case of frozen water pipes, a longitudinal split occurs parallel to the axis of the pipe.

The same phenomenon occurs in other cylindrical pressure vessels because of the generation of large hoop stresses. If we consider a borehole as a pressure vessel, we would expect fractures to orient themselves parallel to the axis of the borehole. The major difference between pressurizing a borehole and pressurizing a water pipe is rate of loading. A borehole is pressurized almost instantly and, therefore, does not fail at one weakest point along the borehole wall. Instead, it will simultaneously fail in many locations. Each resulting fracture will be oriented parallel to the axis of the borehole. Failure by this mechanism has been recognized for many years and is commonly called radial cracking (Fig. 2.8).

Direction and extent of the radial crack system can be controlled by the

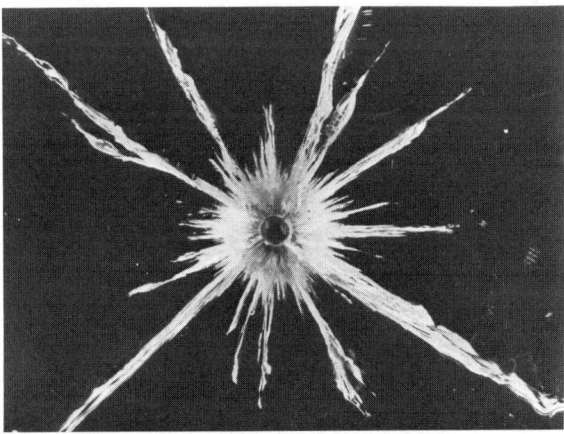

Figure 2.8   Radial cracking in plexiglas.

selection of the proper *burden*, the distance from the borehole to the face (Fig. 2.9). The larger the burden on the borehole, the more resistance to motion therefore the radial cracks extend to a greater distance behind the borehole.

The second major breakage mechanism occurs after the radial cracking has been completed. There is a time lag before the second breakage mechanism goes into play. The second mechanism influences the breakage perpendicular to the axis of the charge.

Before the second breakage mechanism is discussed, let us form a mental picture of what has happened during the radial cracking process. Stress wave (shock) energy has caused minor cracking or microfracturing on the borehole walls and at discontinuities between the hole and the face. The distance from the hole to the face or relief is called the burden. The sustained gas pressure, which follows the shock pressure, puts the borehole walls into tension because of the hoop stresses generated and causes the existing microfractures to grow. The high-pressure gases extend fractures throughout the burden. The burden in massive rock is transformed from a solid rock mass into one that is broken by the radial cracks in many wedge-shaped or pie-shaped pieces. These wedges function as columns, supporting the burden weight. Columns become weaker if their length to diameter ratio or slenderness ratio increases. Therefore, once the massive burden is transformed into pie-shaped pieces with a fixed bench height, it has been severely weakened because of the fact that its slenderness ratio has increased.

The work process has not yet been completed since the expanding borehole still contains high-pressure gases. These gases move into the radial crack network and subject the wedges of rock to forces acting perpendicular to the axis of the hole. One can say they are pushing towards relief or towards the line of least resistance. This concept of relief perpendicular to the axis of the hole has been known for well over a hundred years. Relief must be available perpendicular to the axis of the hole for borehole charges to function properly. If relief is not available, only radial cracks will form and boreholes will crater or the stemming will be blown out. In either case, the fragmentation suffers and environmental problems result.

In most blasting operations, the first visible movement occurs when the face

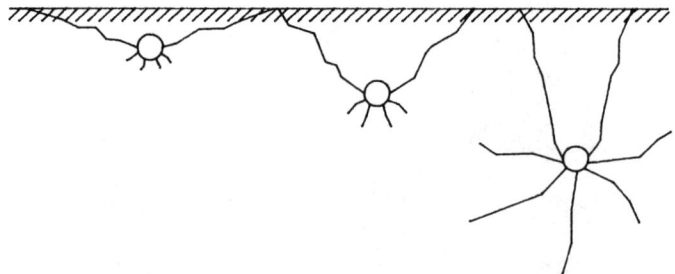

Figure 2.9   Influence of confinement on radial crack network (plan view).

bows outward near the center. In other words, the center portion of the face is moving faster than the top or bottom of the burden (Fig. 2.10).

This type of bowing or bending action, however, does not always occur. You can find cases where instead of the center bowing outward, the top or bottom portion of the burden is cantilevering outward (Fig. 2.11).

Burden stiffness. Differential movement of the face causes the burden to break in the third dimension. This breakage mechanism is called *flexural failure*. To properly discuss flexural failure, one must realize that these individual pie-shaped columns of rock caused by the radial cracking will also be influenced by a force perpendicular to the length of the column. This would be similar to beam loading conditions. When placing a load on a beam, the stiffness ratio controls the stress level at which failure will occur. The stiffness ratio relates the thickness of the beam to its length. The effect of the stiffness can be explained by using, as an example, a full-length pencil. It is quite easy to break a pencil with the force exerted with your fingers. However, if the same force is exerted on a two-inch long pencil, it becomes more difficult to break. The pencil's diameter hasn't changed, nor have the material's properties, the only thing that has changed is its length. A similar stiffness phenomenon also occurs in blasting. The burden rock is difficult to break by flexural failure when bench heights are near equal or less than the burden dimension. When bench heights are many times the burden dimension in length, the burden rock is easily broken.

Two general modes of flexural failure of the burden exist. In one case, the burden bends outward or bulges in the center more quickly than it does on the top or bottom. In the second case, the top or the bottom of the burden moves at a higher rate than the center. When the burden rock bulges at its center, tensile stresses result at the face and compression results near the charge. Under this type of bending condition, the rock will crack from the face toward the hole (Fig. 2.10). This mode of failure generally leads to desirable breakage.

In the second case, the rock is cantilevered outward (Figure 2.11) and the face is put into compression and the borehole walls are in tension.

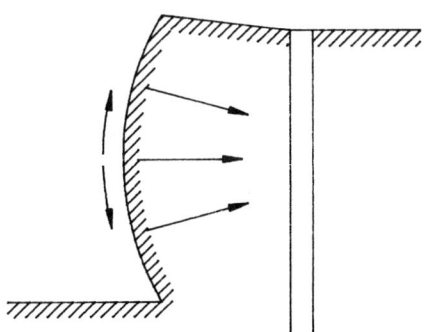

Figure 2.10   Normal face displacement.

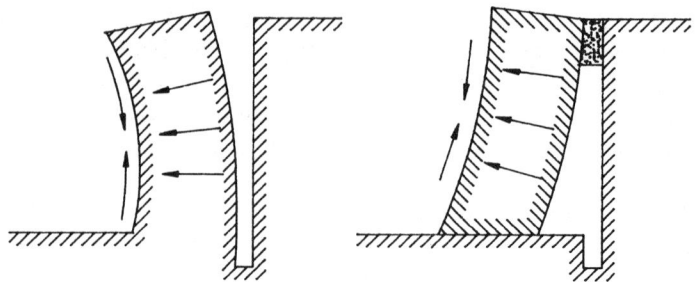

Figure 2.11   Cantilever bending of face.

This second case is undesirable. This mechanism occurs when cracks between blastholes link before the burden is fully broken and is normally caused by insufficient blasthole spacings. When the cracks between holes reach the surface, gases can be prematurely vented before they have accomplished all potential work. Airblast and flyrock can result along with potential problems at the grade or floor level.

The bending mechanism or flexural failure is controlled by selecting the proper blasthole spacing. As bench heights decrease, you must change the spacing distance between holes to overcome the problems of increased stiffness.

Stiffness model.   In order to better understand the rock breakage process, Haghighi (1985) has developed a finite element model that closely resembles the radial crack network before rock movement occurs. The model was unique in that it allowed the study of radial cracks, which were partially pressurized. An interim technique was used to recalculate and update the borehole pressures as borehole volume increased. The model was designed to study two important aspects of bench blasting. The first model was used to determine the stiffness and effect of the bench height on flexural failure and the second was to determine the effect of changing geologic conditions and rock strengths on the movement of the burden itself (Fig. 2.12).

Model parameters were chosen so that actual burden displacement could be predicted and compared with that of actual field results. The model consisted of a single hole, four inches in diameter. The burden was ten feet, collar stemming and subdrilling below grade were eight feet and four feet respectively and the bench height was varied from twelve to one hundred feet. Therefore, the stiffness ratio (ratio of bench height divided by burden) changed from 1.2 to 10. The explosive parameters used in the model were those of ammonium nitrate and fuel oil. The borehole was initially pressurized with explosive gases to 425,000 psi.

Stiffness analysis.   The burden (B) was held constant at 10 feet throughout the analysis, therefore, changes in the stiffness resulted from varying the bench height (L). As the L/B ratio increased from 1.2 to 10, the bench height increased

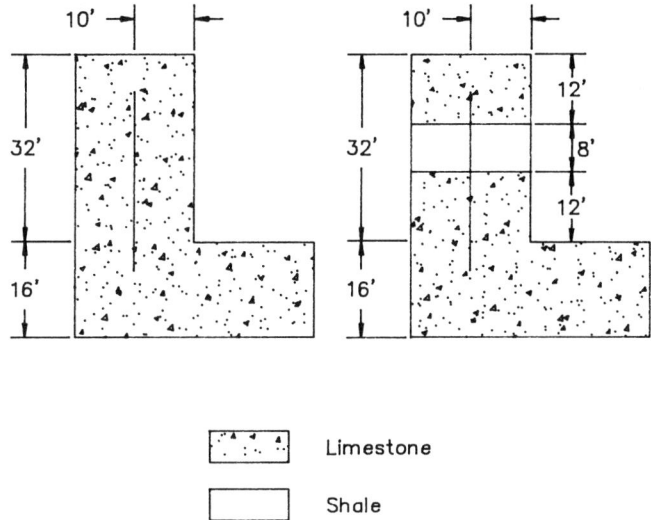

Figure 2.12 Finite element model configuration to test geologic influence.
(Haghighi 1986)

from 12 to 100 feet. Face displacements and outer fiber stresses were calculated for different locations on the rock face.

For discussion purposes, we will focus on four different L/B ratios, 1.2, 2.4, 3.6, and 4.0. With an L/B ratio of 1.2, there was little displacement on the face of the shot; instead, local crushing occurred around the hole (Fig. 2.13). As the bench height increased and the L/B ratio became 2.4 less, local crushing occurred and the model indicated a maximum displacement on the face, of 43 inches. For an L/B ratio of 3.6, the model indicated a displacement of 186 inches, and at an L/B ratio of 4.0, the displacement was 279 inches or 23.2 feet. Figure 2.14 shows both the deformed geometry configuration of this model and scaled displacements.

Further analysis was conducted applying beam bending theory in an effort to quantitatively explain the behavior of burden rock subjected to explosive gas pressure loads. The graph of L/B ratio versus displacement gives a better understanding of why this ratio is so important in bench blasting (Fig. 2.15). The graph shows that the displacement increased at a slower rate when L/B ratios varied between 1 and 3.5. At 3.5, there was a distinct change in slope of the curve, which indicated that for the same gas pressure there was additional displacement. When the L/B ratio was greater than 6, there were significant increases in displacement with small changes in the L/B ratio. Research conducted by Konya (1968) and empirical data from the field, indicates that a practical upper limit for L/B ratio is approximately 4. Above 4, the effects of stiffness are minimal and are no longer considered in field design. It is interesting to note that in this finite element analysis the curve made two major slope changes, one at approximately 3.5 and the other at 6. What is

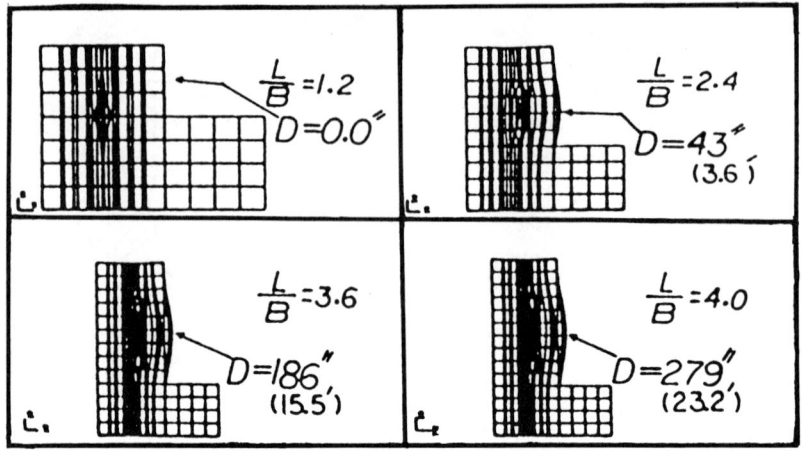

Figure 2.13   XZ-view of the deformed geometry configuration as L/B ratio changes from 1.2 to 4.0. (Haghighi 1985)

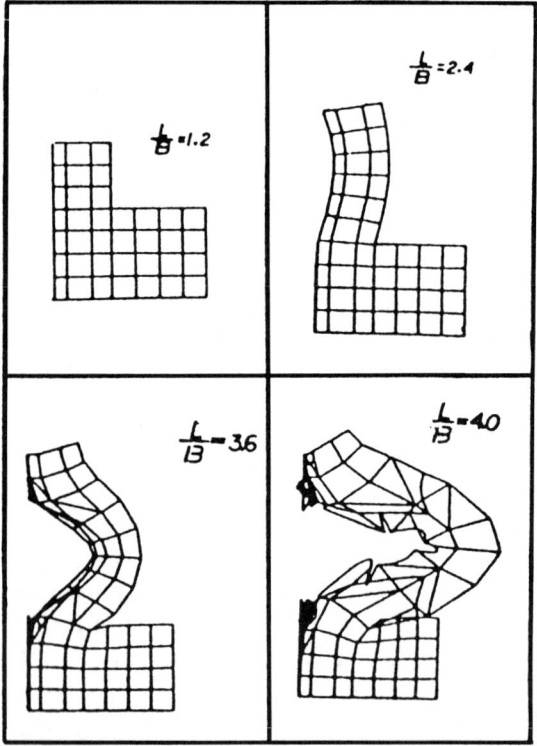

Figure 2.14   Actual displacements for XZ-view as L/B changes from 1.2 to 4.0 (no relative displacements). (Haghighi 1985)

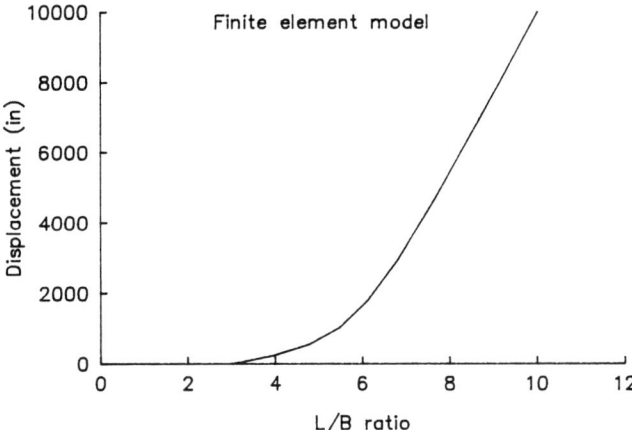

Figure 2.15   L/B ratio versus deflection for the finite element model. (Haghighi 1985)

important to note is that in both of these transitional zones, there was a significant increase in displacement for small changes in the L/B ratio.

Geological Effects on Displacement

In order to demonstrate the significance of beds of different materials on bench blasting, different models were analyzed using the same finite element approach (Fig. 2.16). In one example, two materials were considered, a limestone rock with a modulus of elasticity of 5.0 times 10^6 and a poisson's ratio of 0.21. A second material was considered which was a weak shale having a modulus of elasticity of 0.5 times 10^6 and a poisson's ration of 0.20. A finite element analysis was conducted as previously described. The difference in this model was that the L/B ratio was held at 4. All parameters were held constant except the composition of the bench itself. Figure 2.16 shows a shale layer located between two layers of limestone, representing a condition in which the soft layer is present in the harder rock bench. The analysis indicated that the middle of the soft layer at the free face could be displaced as much as 45 feet.

Where the bench consisted of homogeneous limestone, the model indicated a displacement of 6.7 feet.

High-speed photography has proven that weak materials interbedded in massive hard rock do blow out in a similar fashion as demonstrated by this finite element analysis leaving the massive rock poorly broken.

Field considerations.   For unconfined charges, the shock pressure produces the majority of the breakage. When blasting with boreholes, there are three mechanisms of rock breakage: the minor microfracturing caused by the shock energy, the major radial fracturing, and the breakage perpendicular to the axis of the

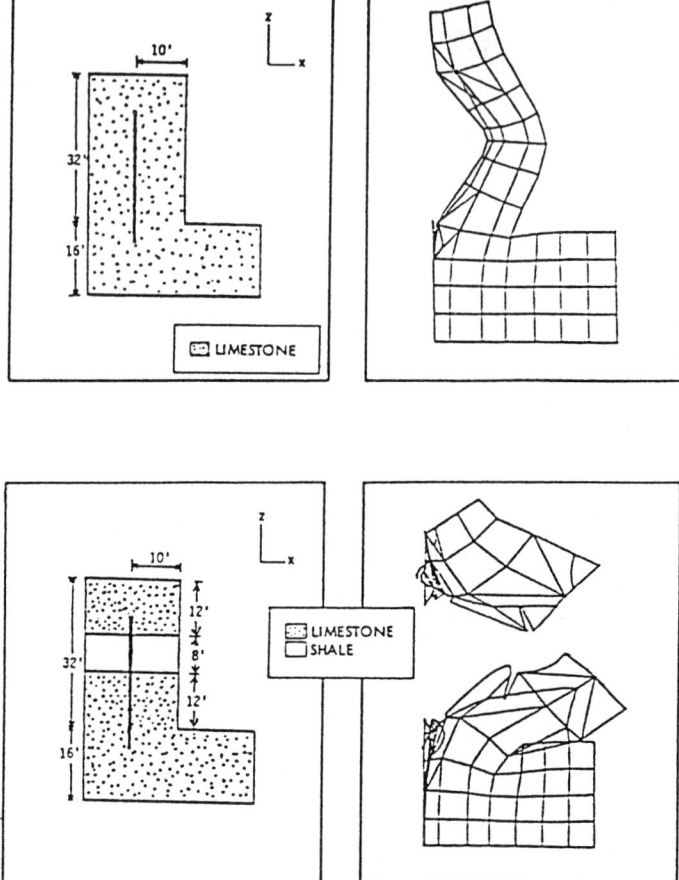

Figure 2.16   Changes in displacement as a result of weak rock layer.

charge by flexural failure are caused by the gas pressure. One practical example follows to illustrate the role of both the shock and gas pressure in rock fragmentation.

If the rock bench to be blasted contains a large mud seam which intersects both the face and the borehole, totally different fragmentation would result than if the rock contained no mud seam (Fig. 2.17).

When a large mud seam exists and no stemming is used across the seam to confine the gases, little rock fragmentation results. The mud seam would not significantly affect the minor breakage which results from the shock energy. It would, however, allow premature venting of the gases and drastically reduce the gas energy necessary for both radial cracking and flexural failure.

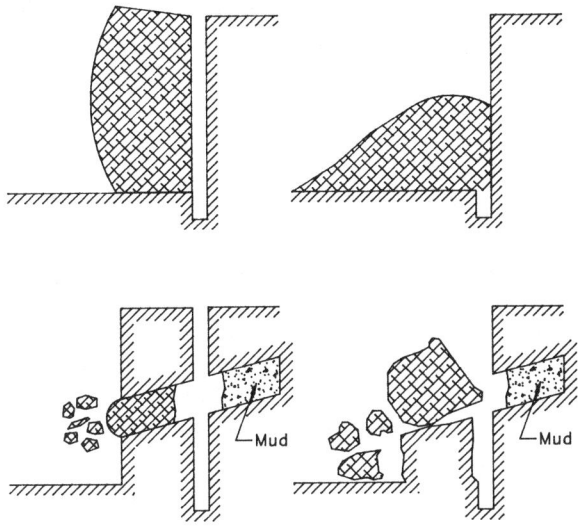

Figure 2.17   Effect of mud seams in field blasting.

Monitoring Delivered Energy

The amount of energy actually delivered in the blasthole controls rock break-age as well as environmental effects. Insufficient energy causes bouldering, vio-lence, and high-vibration levels. It is commonly assumed that the rated theoretical energy of the explosive is delivered in the blasthole. That assumption can often be in error. There are many factors that control the percentage of useful energy deliv-ered. Low energy levels can result if the explosive is improperly mixed, or im-properly used. The local environment of the blasthole may also cause low energy release. When blasts are not breaking properly, we have traditionally thought that this was a result of energy loss due to geologic conditions. With new monitoring techniques, however, we have found that commonly the geology was not the prob-lem, instead it was insufficient energy release.

Monitoring energy release in the field in a production environment has been difficult because it required sophisticated electronic equipment which was not readi-ly available to the operator. In 1988, equipment was developed which was field portable, did not interfere with production, and would produce reliable results which would give the operator information on energy release within the blasthole.

Two types of measurements can be made in the field. Detonation velocity can be measured as well as borehole pressure. These two measurements quantify the amount of energy delivered and can be compared to that anticipated under ideal conditions. This provides the operator information to adjust subsequent blasts and modify his design based on actual delivered energy instead of relying on theoretical energy. It is no longer necessary to guess at the cause of problems since hard data can be generated easily and inexpensively.

PROBLEMS

1. What is the controlling factor which influences the direction of radial crack growth?
2. The stress wave can cause microfractures on the borehole walls. Why can't the stress wave propagate the microfractures to produce large radial cracks?
3. The radial cracks are oriented in what specific direction with the borehole axis? What type of stress causes them to be oriented in this direction?
4. Why does breakage in plaster shooting result from the shock component of energy rather than the gas component?
5. How does flexural failure play a role in the breakage process?
6. Why is cantilever bending of the burden undesirable?
7. Which blasting dimension is used to help compensate for low stiffness ratios?
8. Why are mud seams a concern in blasting and what effect do they have?
9. Are there methods for monitoring the explosives energy delivered in the borehole? If so, what is measured?

REFERENCES

ASH, R.L. and KONYA, C.J., "Flexural Rupture: A New Theory on Rock Breakage by Blasting," in *Proceedings of the International Conference on Explosives and Blasting Technique,* Linz, Austria: WIFI, 1975, pp. 13–19.

HAGHIGHI, R.G. and KONYA, C.J., "The Effect of Bench Movement with Changing Blasthole Length," *Proceedings of the First Mini-Symposium on Explosives and Blasting Research,* Society of Explosives Engineers, Montville, OH 1985.

HAGHIGHI, R.G. and KONYA, C.J., "Effects of Geology on Burden Displacement," *Proceedings of the Twelfth Conference on Explosives and Blasting Techniques,* Society of Explosives Engineers, Montville, OH, 1986.

KONYA, C.J., "Spacing of Explosives Charges," M.S. thesis, University of Missouri, Rolla, 1968.

KONYA, C.J., "The Mechanics of Rock Breakage Around a Confined and Air Gapped Charge," *Proceedings of the International-Blasting Section of the Scientific Society of Buildings,* 1973.

OTUONYE, F.O., SKIDMORE, D.R., and KONYA, C.J., "Measurements and Predictions of Borehole Pressure Variation in Model Blasting Systems," *Conference Proceedings of the First International Symposium on Rock Fragmentation by Blasting,* Lulea, Sweden, August 22–25, 1983.

3

Seismic Energy and Vibration Considerations

The previous chapter discussed the manner in which the useful energy that results from an explosive reaction breaks rock. An explosive reaction, however, does not produce only energy which causes rock to break. When energy levels are below those necessary to cause plastic deformation or failure and are within the elastic limits of the material, no breakage results. Stresses produced within the elastic limits occur at two different times within a blast. As pressures are building in the hole during the chemical reaction, the rock is first stressed below its elastic limits. When all fracturing has ceased, because energy levels have fallen and energy has been used in the breakage process, stresses will remain which are within the elastic limits of the material. Stress levels within the elastic limits do not cause useful work to be performed, however, they do generate seismic waves.

If blasts are properly designed, the amount of seismic energy can be reduced since more energy will be used in rock breakage. However, if blasts are improperly designed, seismic energy increases. The seismic energy is commonly considered waste energy, because it serves no useful purpose. This waste energy causes the vibration problem in blasting. Surface blasting operations near residential areas can cause annoyance and possible damage to structures. For this reason, in the design of any blast, you must consider the effects of this waste energy. Seismic energy levels must be predictable and reproducible. Seismic monitoring must be fully understood by the explosives engineer and must be accurate. It does no good to have seismic records which are poorly taken and neither stand up in litigation nor can be used for engineering purposes. In that case, the expense of taking the records was wasted.

Wave parameters. Wave motion is described by its fundamental properties called wave parameters. These are the properties measured and, as a result, have numbers or values assigned to them when discussing wave motion or vibration. A simple harmonic wave motion is illustrated in Fig. 3.1 and represented by the equation.

$$y = A \sin (wt) \tag{3.1}$$

where:

y = displacement at any time t, measured from the zero line or time axis
t = time
A = amplitude or maximum value of y
$w = 2\pi f$
T = period or time for one complete oscillation
f = frequency, the number of vibrations or oscillations occurring in one second, designated Hertz, Hz

Period and frequency are reciprocals so that

$$f = 1/T \quad \text{or} \quad T = 1/f \tag{3.2}$$

Wave length L is the distance from any point on a wave to the same point on the next cycle of the wave. It is usually measured from crest to crest or trough to trough with the length expressed in feet. It is equal to the wave period T multiplied by the propagation velocity V.

$$L = VT \tag{3.3}$$

The seismic sensor.   Vibration instrumentation is designed to measure and record the motion of the vibrating earth and if it fails to do this, it is worthless. The vibration record must be a faithful reproduction of the earth motion. The instrument which does this is called a seismograph and consists of two parts, a sensor and recorder.

The sensor case actually encloses three independent sensor units placed at

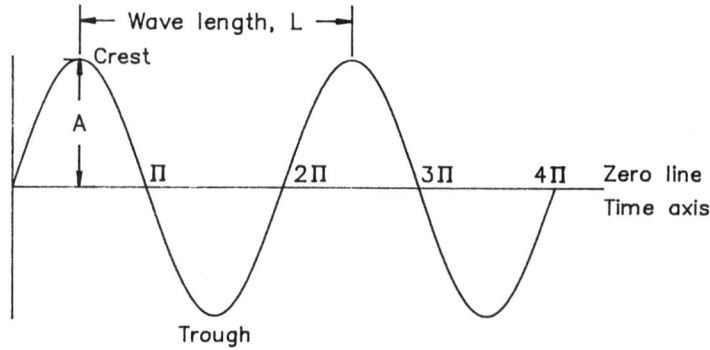

Figure 3.1 Wave motion and parameters.

right angles to each other. Two units lie in the horizontal plane at right angles to each other while the third unit is oriented in the vertical plane. Each sensor can respond only to motion along its axis. The three sensors each provide a vector component of the ground motion. The three units are enclosed in a case as shown in Fig. 3.2.

The sensor is a voltage or signal generator. The earth moves and it responds. The sensor is usually an electromagnetic transducer which converts ground motion into electrical voltage. It operates on the principle that if a conductor moves in a magnetic field it will generate an electrical potential. A conductor in the form of a coil of wire is suspended in a permanent magnet field. The magnet is attached to the sensor case and cannot move, while the coil is suspended in the magnetic field by springs or hinges and is free to move. Any movement of the coil relative to the magnetic field will generate an electrical voltage. When the ground vibrates, the sensor case and attached magnet will also vibrate, but the inertia of the coil will cause it to remain motionless. Relative motion between the coil and the magnetic field results, and a voltage is generated. The voltage is proportional to the speed of the relative motion. If the earth vibrates slowly, a small voltage is generated. If the earth vibrates rapidly, a large voltage is generated. Once the coil starts moving, it will naturally tend to continue moving or vibrating but the damping of the instrument prevents this and stops the coil from vibrating freely.

A schematic diagram of the sensor transducer to shown in Fig. 3.3. The voltage output of the sensor goes to the recorder where it is converted back into motion. This motion is recorded photographically or on a chart recorder and is referred to as an analog record of the ground motion. The record will have three lines or traces, one for each sensor unit.

A galvanometer or equivalent system converts the sensor voltage back into motion, using the reverse of the voltage generating principle. Voltage generated at the sensor will flow, as a current into the galvanometer circuit causing the galvanometer coil to move. If amplification of the ground motion is called for, it is done at this point. Timing lines and calibration signals are also put on the record.

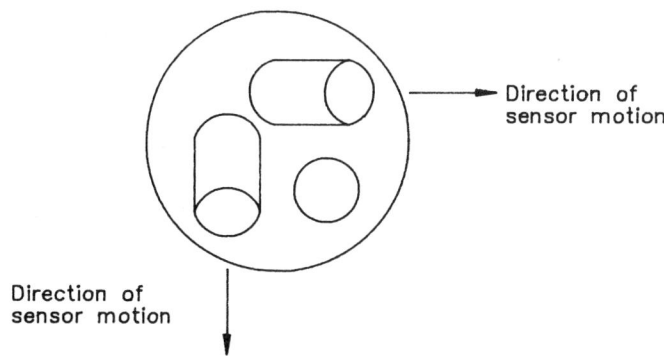

Figure 3.2   Seismograph sensor.

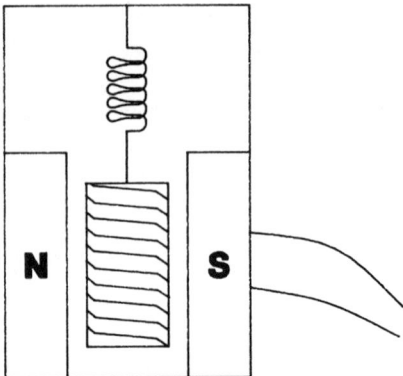

Figure 3.3 Sensor mechanism.

There are also seismographs that record on magnetic tape requiring a playback system and a chart recorder to obtain a record. Although somewhat more involved, this type has many advantages. Various amplifications of the ground motion can be used in the playback and various parameters of the ground motion can be studied using appropriate analysis techniques. In the direct recording technique, once the shot is fired and recorded, there is only a static hard copy record. On the tape the blast lives forever and can be recalled at anytime and as often as needed—it is an instant replay. A single tape can record many blasts and tapes are inexpensive and can be purchased in many different types of stores.

Types of seismograph systems.   Many types of seismographs are available today. Each performs the basic function of measuring ground motion but supplies much additional information. The variations are a response to advancing technology in both the blasting and vibration fields which has revealed a need e.g., frequency analysis, and a constraint, e.g., reduce cap firing spread. The industry is responding to a time of expansive growth. The principal types of seismograph are described.

> Analog seismograph: a three-component system that produces a record of the ground motion. It is called analog because the record is an exact reproduction of the ground motion only changed in size—amplified, or de-amplified.
>
> Tape seismograph: the same as the analog seismograph, except that it records on a magnetic tape cassette instead of directly producing a record. A record of the ground motion is obtained by use of a playback system and a chart recorder.
>
> Digital seismograph: the voltage signal generated by the sensor unit is digitized, that is, the incoming signal voltage is sampled every millisecond, therefore, the data is contained in 1000 voltage values in each second of motion. A hard copy record of the blast vibration can be made from the digitized data at the site.

Vector sum seismograph: the standard seismograph system consists of three mutually perpendicular vector components of the ground motion. The resultant ground motion can be determined by combining the vector components using the following relationship:

$$S = (V^2 + L^2 + T^2)^{0.5} \qquad (3.4)$$

where:

S = resultant motion
V = vertical component of motion
L = longitudinal component of motion
T = transverse component of motion

This calculation is done electronically by the vector sum seismograph. The calculation is continuous so that the resultant is a vector recomposition of the ground motion over the complete time domain.

Triggered seismograph: any seismograph which automatically starts to record when the ground vibration level reaches a predetermined set value, which triggers the system.

Bar graph seismograph: a three-component system that records at a very slow speed. It can be put in place and left to record for periods of up to thirty or sixty days. It doesn't record the wave forms. It only records the maximum ground motion of the three components as a deflection on the record which has the appearance of a bar graph.

Most seismographs are equipped with meters that register and hold the maximum value of the vibration components and the sound level. Other seismographs are equipped to produce a printout which gives a variety of information such as maximum value for each component, frequency of vibration for the maximum value, maximum displacement, maximum acceleration, vector sum, and sound level. Blast information such as date, blast number, time, location, job designation, and other pertinent information can also be added to the printout.

Vibration parameters.   Wave parameters were discussed earlier. Vibration parameters refer to the character of the ground motion. They are displacement, velocity, and acceleration. When the ground vibrates because of the passage of a seismic wave, the rock particles move or are displaced from an equilibrium position. This is called displacement. How fast the particle moves, is its velocity. It can also exert force that is proportional to the particle's acceleration which is the rate of change of velocity. These fundamental vibration parameters are defined as follows:

Displacement   The distance that a rock particle moves from its equilibrium position. It is measured in fractions of an inch, usually thousandths.

Velocity | The speed at which the rock particle moves when it leaves its rest position. Particle velocity is measured in inches per second.

Acceleration | The rate at which particle velocity changes. Force exerted by vibrating particle is proportional to the particle acceleration. Acceleration of gravity is measured in fractions of "g" ($g = 32.2$ ft/sec^2).

The equation for these parameter are the following with the maximum values specified:

	Standard	Maximum	
Displacement	$Y = A \sin(wt)$	$Y = A$	(3.5)
Velocity	$V = AW \cos(wt)$	$V = AW$	(3.6)
Acceleration	$a = -AW^2 \sin(wt)$	$a = -AW^2$	(3.7)

Particle velocity is considered the best descriptor of damage and the standards of damage are based on particle velocity. Vibration seismographs, therefore, measure particle velocity. In addition, there are displacement seismographs and acceleration seismographs. It is not necessary to have separate systems since a velocity seismograph can be equipped to electronically integrate or differentiate the velocity signals to produce displacement or acceleration data. Some seismographs do this routinely for each shot and present the data on the printout.

Vibration Records and Interpretation

Normally a seismograph record will show the following:

Three lines or traces, one for each vibration component. A fourth line or trace for the acoustic or sound level.

A calibration signal for each trace

Timing lines which appear as vertical lines running across all or part of the record.

An example of a typical seismogram, or vibration record, is shown in Fig. 3.4.

The seismograph polarizes ground motion into three perpendicular components as shown in Fig. 3.5. The components of motion are usually designated, vertical, longitudinal, and transverse.

Vertical: | motion up and down, designated V
Longitudinal or radial: | motion along a line from the source to the recording point, designated L or R

Transverse: motion at right angles to a line from the source and
                               the recording point, designated T

An arrow inscribed on the top of the sensor should be pointed at the shot.
Then vibration traces will consistently occur in the same sequence, with the arrow
indicating the L component. The direction of motion will then be consistent from
shot to shot.

Each trace represents how the ground is vibrating in that component or direction. In the case of a velocity seismograph (the usual case), each trace shows how
the particle velocity is changing from instant to instant.

Similarly, if the seismograph is a displacement system or an acceleration
system, the traces will show the instant to instant change in these parameters. The
acoustic trace shows how the sound level changes with time.

Record reading and interpretation.   With the introduction of computer
systems and processing to vibration seismographs, much of the analysis appears on
a printout. A flood of information has been forthcoming. The printout also produces
a hardcopy wave-form record. Analysis of this wave-form record will be discussed.
The maximum vibration level, that is the peak particle velocity, is of primary
interest. The analyst studies the record and selects the largest amplitude, measured
either up or down from the zero line on each given trace. This value in inches

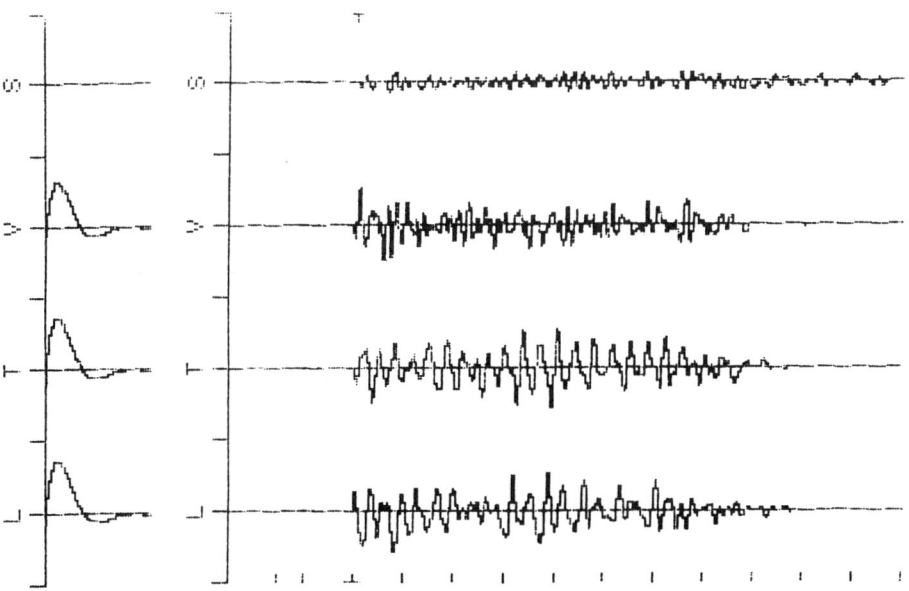

Figure 3.4   Vibration record.

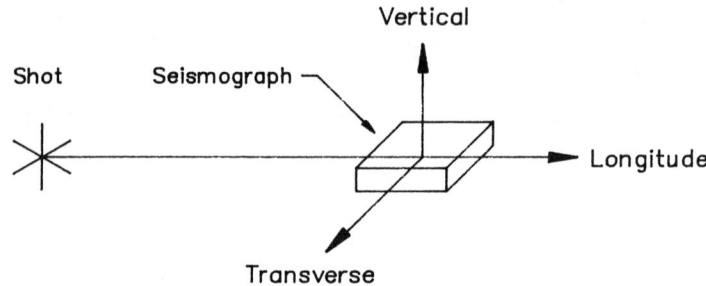

Figure 3.5 Vibration components.

(usually fractions of an inch, e.g., 0.31) is then divided by the instrument gain setting. The result is the maximum ground motion. Velocity seismographs normally have a fixed gain of one, but in special cases amplification or deamplification may be used. The seismograph trace motions are complex rather than sinusoidal and are not usually symmetrical, hence, the maximum single amplitude is measured in order to determine the maximum ground motions. Figure 3.6 illustrates the basic measurement techniques.

$$\text{Trace amplitude } A = 0.31$$
$$\text{Seismograph gain} = 2$$
$$\text{Ground vibration} = \text{trace amplitude/seismograph gain} \qquad (3.8)$$

$$V = A/gn = \frac{0.31}{2}$$
$$V = 0.155$$

For a velocity seismograph, this value represents a peak particle velocity of 0.155 inches per second. For a displacement seismograph, this value represents a peak ground displacement of 0.155 inches. For an acceleration seismograph, this

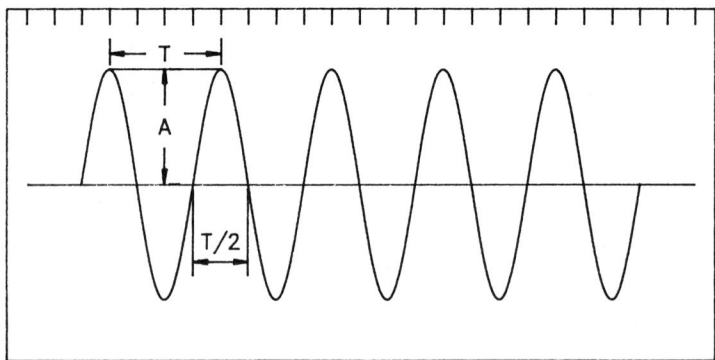

Figure 3.6 Measurement of vibration amplitude and period.

value represents a peak ground acceleration of 0.155 g. These are merely numerical illustrations and have no other significance.

Frequency measurement on a vibration record, at best, is imprecise because of the complex motion on the record. A sinusoidal wave would have equal amplitudes for the crests and trough, and would have equal distance between successive crests or troughs. This is an ideal condition that normally does not occur. The measurement procedure must be adjusted to fit the circumstances. Such adjustments usually add to the lack of precision and increase the measurement error. The procedure will be illustrated for an ideal case and for a more normal case.

Figure 3.6 shows a sinusoidal wave form with equal successive crests and troughs. Select two successive crests and count the number of timing spaces between them. There are four, since each timing space has a value of .02 seconds, then:

$$T = 4 \times .02$$
$$T = .08 \text{ sec}$$

since

$$f = \frac{1}{T}$$
$$f = \frac{1}{.08}$$
$$f = 12.5 \text{ Hz}$$

The half period of the wave can be measured by noting the points of zero crossing of the time axis. This is also illustrated in Fig. 3.6. There are two timing spaces which represents .04 seconds, hence

$$T/2 = .04$$
$$T = .08$$
$$\text{and } f = 12.5$$

yielding the same result for the ideal case. So much for a sinusoidal wave. Next consider a non-sinusoidal wave as illustrated in Fig. 3.7.

The successive peaks or troughs are not of equal amplitude. Also, the time spacing from crest to crest, trough to trough, or crest to trough or between zero crossings is different. This is a judgment call by the analyst, and as such is error prone. First consider measuring from the lowest trough A to the following highest peak C. This is a half period and the analyst must estimate between timing lines. There are 3.3 timing spaces from the trough to the crest, since each timing line represents 0.02 seconds.

$$\frac{T}{2} = 3.3 \times 0.02$$
$$\frac{T}{2} = 0.066$$
$$T = 0.132 \text{ seconds}$$

and

$$f = \frac{1}{T}$$
$$f = \frac{1}{0.132}$$
$$f = 7.6 \text{ Hz}$$

Now consider two other estimates of the period and frequency. Measure the half period from the highest crest C to the following trough E. There are 4.8 timing spaces which is .096 seconds.

$$T/2 = .096$$
$$T = .192 \text{ seconds}$$
$$f = 5.2 \text{ Hz}$$

Now, measure the zero crossings B to D for the highest peak. There are 5.2 timing spaces which is .104 seconds.

$$T/2 = 0.104$$
$$T = 0.208 \text{ seconds}$$
$$f = 4.8 \text{ Hz}$$

The three measurements give three different values for the frequency 7.6 Hz, 5.2 Hz and 4.8 Hz. The 4.8 Hz measured from the zero crossings is the preferred value.

Different values for the wave frequency will result from the measurement of full-wave oscillation and half-wave oscillation. Also in practice, it is usually necessary to measure fractions of a timing space which introduces error into the measurement because of the inaccuracy inherent in estimation.

Frequency estimated by reading from the record is highly susceptible to error. Frequency measured electronically by the seismograph system has good accuracy

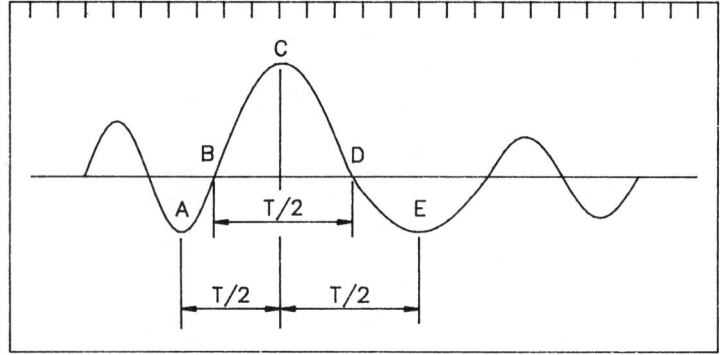

Figure 3.7 Half-period measurement.

and repeatability. The large scatter in data characteristic of early investigations and research no doubt is due at least in part to the difficulties associated with measuring frequency on a record.

Acoustic levels usually employ a base level and the instrument responds only above this level. This base-level value is then added to the computed value to give the true sound level reading. The following example illustrates the procedure:

sound level =
    (acoustic trace amplitude/acoustic gain) + acoustic base level   (3.9)

$$
\begin{aligned}
\text{acoustic trace amplitude} &= 0.180 \text{ inches} \\
\text{acoustic gain} &= 0.005 \\
\text{acoustic base level} &= 80 \text{ dB} \\
\text{sound level} &= \frac{(0.180}{0.005)} + 80 \\
&= 36 + 80 \\
&= 116 \text{ dB}
\end{aligned}
$$

Human ear sensitivity will only detect sound level change of about 3 dB. Less than 3 dB generally is not noticed. Electronic measurement and computer analysis have been very beneficial for the vibration industry. Problems that seemed almost insurmountable a few years back have now been made subject to analysis. The ease with which things can now be done may be a sort of mirage. The present state of the art was arrived at because people perceived problems and set about solving them. A wave form record is a statement of problems asking to be looked at. The printout of numbers and values so commonly produced by todays seismographs does not reveal the unsolved problems hidden in the shaking earth and responding structures. A vibrating structure whose natural frequency happens to match one of the frequencies in an incoming seismic wave will respond to that frequency with an amplified motion. What about the mismatched frequencies? These will produce a forced motion in which the structure is driven by the vibration. The resonant matching frequency may not be obvious to the analyst examining a wave-form record so he needs the printout which shows the frequency. But the print out reveals only the maximum. A harmonious blending of the two approaches will probably best serve the industry.

Principle Factors Affecting Vibration

When a blast is fired, the vibration level is controlled by two principal factors, distance and charge size. Obviously, it is safer to be far away from a blast than to be near it. Equally obvious is that a large explosive charge is more dangerous than a small charge.

Charge—distance relationship. The U.S. Bureau of Mines Bulletin 656 (Nichols, Johnson, and Duvall, 1971) developed a mathematical model called the propagation law which relates peak particle velocity, charge weight, and distance. The formulation is:

$$V = H \, (D/w^a)^b \tag{3.10}$$

where:
   V = predicted particle velocity in in./sec
   w = maximum explosive charge per delay in lb
   D = distance from shot to sensor measured in 100s of ft
        (e.g., for distance of 500 ft, D = 5)
   H = particle velocity intercept
   a = charge weight exponent
   b = slope factor exponent

The Bureau of Mines empirically determined values for H, a, and b for each of the three components of motion longitudinal, vertical, and transverse. The numerical values for H, a, and B are slightly different for each component. These equations are

$$Vr = 0.052 \, (D/W^{0.512})^{-1.63} \tag{3.11}$$

$$Vv = 0.071 \, (D/W^{0.421})^{-1.74} \tag{3.12}$$

$$Vt = 0.035 \, (D/W^{0.521})^{-1.28} \tag{3.13}$$

Use the following approximations:
$$W^{0.5}$$
$$b = -1.6$$

and express D in feet instead of hundreds of feet. The equation for the longitudinal components with Vr replaced by V becomes

$$V = 100 \, (d/w^{0.5})^{-1.6} \tag{3.14}$$

where

   d = distance in ft, shot to sensor
   w = maximum explosive charge per delay in lb

A similar equation is given in *The DuPont Blaster's Handbook* (E.I. DuPont de Nemours & Co., 1977).

$$V = 160 \, (d/W^{0.5})^{-1.6} \tag{3.15}$$

Estimating particle velocity.   What value for the peak particle velocity is likely to result from the detonation of a given explosive charge at a given distance? This value can be estimated using Eq. 3.14 or the DuPont formula which will give a higher value. This advises caution in the use of the formulas. They do not yield exact values, but serve merely as guides, and only give ballpark figures.

An example of such calculation is as follows:

Using

$$V = 100 \ (d/w^{0.5})^{-1.6}$$

and assuming

$$d = 450 \text{ ft}$$
$$w = 225 \text{ lb per delay}$$

then

$$V = 100 \ (450/225^{0.5})^{-1.6}$$
$$V = 100 \ (30)^{-1.6}$$
$$V = 100/230$$
$$V = 0.435 \text{ in./sec}$$

The Bureau of Mines formula can be expressed in many different forms. It can be expressed in terms of charge weight as

$$W = d^2 \ (V/100)^{1.25}$$

To illustrate a calculation of W, use the values

$$d = 600 \text{ ft}$$
$$V = 0.35 \text{ ips}$$

then

$$W = (600)^2 \ (0.35/100)^{1.25}$$
$$W = 3.6 \times 10^5 \ (8.5 \times 10^{-4})$$
$$W = 306 \text{ lb}$$

The values of "a," "b" and "H" are determined by the blasting environment, rock type, rock layering, thickness of overburden, and many other factors. Variations from area to area will apply. The values of $a = 0.5$ and $b = -1.6$ are generally accepted as workable first approximations until applicable data indicate a change. The value of H, however, is highly variable and is influenced by many factors.

Charge weight, distance effects.   The effect of charge weight and distance, are individually shown in the two graphs presented here. The first is a look at charge weight versus particle velocity, and the second is distance versus particle velocity.

consider

$$V = 100 \, (d/W^{0.5})^{-1.6} \qquad\qquad (3.14)$$

or

$$V = 100 \, (W^{0.8}/d^{1.6}) \qquad\qquad (3.16)$$

Assume that a charge W produces a particle velocity V at a distance d. Now let W vary in multiples $2W$, $3W$, etc., while keeping the distance fixed at d and calculate the relative values of V. The relative values of V are plotted against charge weight in Fig. 3.8. Note that the curve is concaved downward.

Again assume a charge W produces a particle velocity V at a distance d. This time let the distance increase in multiples $2d$, $3d$, etc., while keeping the charge fixed at W and calculate the relative values of V. The relative values of V are plotted against distance in Fig. 3.9. Note the dramatic effect that changing distance has on particle velocity.

While the graphs illustrate the individual effects of charge weight and distance some numerical examples will be helpful. Consider the following:

Question: If the charge weight is doubled, how much will the particle velocity increase?

$$V_1 = 100 \, (d/W^{0.5})^{-1.6}$$
$$V_2 = 100 \, [d/(2W)^{0.5}]^{-1.6}$$
$$V_2 = 100 \, (d/1.4W^{0.5})^{-1.6}$$
$$V_2 = (1.4)^{1.6} \, 100 \, (d/W^{0.5})^{-1.6}$$
$$V_2 = 1.7 \, V_1$$

Answer: Doubling the charge increases the particle velocity 1.7 times. (Note: It is not double.)

Question: If the charge weight is cut in half, how much will the particle velocity decrease?

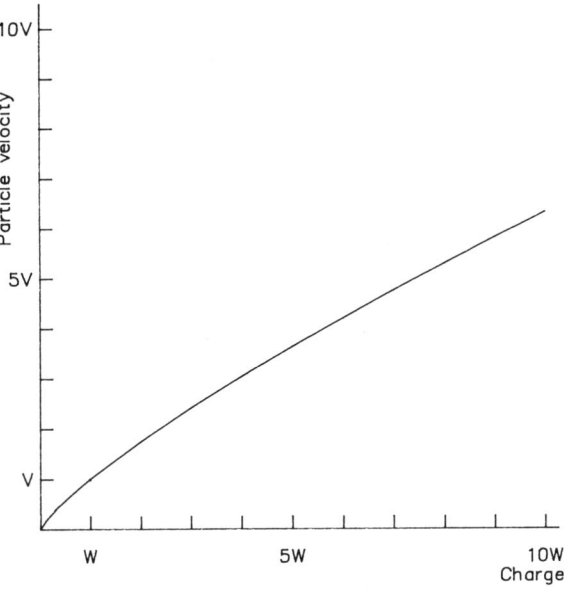

Figure 3.8   Particle velocity—charge weight relationship.

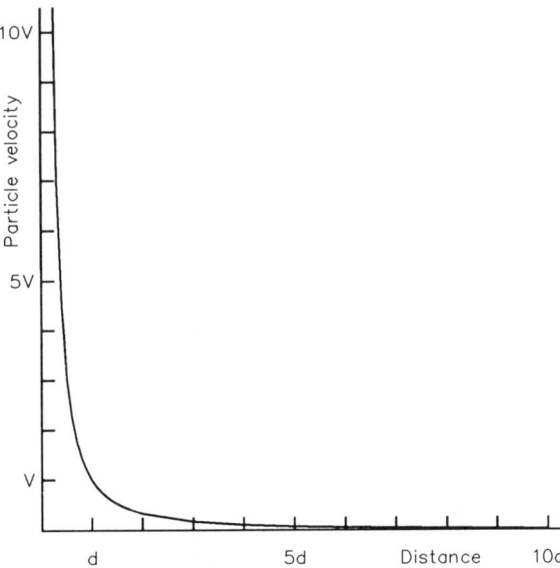

Figure 3.9   Particle velocity—distance relationship.

$$V_1 = 100 \ (d/W^{0.5})^{-1.6}$$
$$V_2 = 100 \ [d/(W/2)^{0.5}]^{-1.6}$$
$$V_2 = -0.59 \ V_1$$
$$V_2 = 0.6 \ V_1$$

Answer: Cutting the charge in half will decrease the particle velocity to six-tenths its original value. (Note: It is not cut in half.)

Question: If the distance is doubled, how much will the particle velocity decrease?

$$V_1 = 100 \ (d/W^{0.5})^{-1.6}$$
$$V_2 = 100 \ (2d/W^{0.5})^{-1.6}$$
$$V_2 = (1/2^{1.6}) \ V_1$$
$$V_2 = 0.33 \ V_1$$

Answer: If the distance is doubled, the particle velocity is reduced to one-third of its original value. (Note: It is reduced more than half.)

Question: If the distance is cut in half, how much will the particle velocity increase?

$$V_1 = 100 \ (d/W^{0.5})^{-1.6}$$
$$V_2 = 100 \ (d/2/W^{0.5})^{-1.6}$$
$$V_2 = 3.03 \ V_1$$

Answer: If the distance is cut in half, the particle velocity will be tripled. (Note: It is increased much more than doubled.)

The last example is particularly revealing of the increase in particle velocity when distance decreases. For example, consider a highway construction project with a house 500 ft away from the blasting. The blasting, however, is moving down the road and at its nearest approach to the house will be 250 ft away, half the present distance. If the present blasting generates a peak particle velocity of 0.8 in. per sec it will increase and be of the order of 2.4 in. per sec at 250 ft from the house. Obviously this tells the operator to adjust his blasting procedure.

The same scenario can be applied to a quarry working two faces, one close to homes, the other farther away. Or a strip mine working several pits, one close to homes, the others farther away.

The value of H was previously indicated to be highly variable. The calculated values are ratios and are not affected by the value of H which cancels out just as the fixed charge value W does.

$$V_1 = H \, (d_1/W^{0.5})^{-1.6}$$
$$V_2 = H \, (d_2/W^{0.5})^{-1.6}$$
$$V_1/V_2 = (d_1/d_2)^{-1.6}$$

Shooting in the same area the values of "a" and "b" should remain relatively constant.

Vibration control.   In field operations, it is essential to know and to be able to control what will happen when a blast is fired. Effective control can be achieved through the use of the formulas discussed.

Delay blasting.   As a starting point, delay blasting will be explained. This is a technique whereby a large blast is reduced to a series of small blasts. This is possible through the use of delay caps—particularly millisecond delays. To illustrate, if a charge W is detonated using five delays the effective vibration generating charge is only one-fifth of W.

Consider the example

A shot consists of 20 holes, 320 lbs of explosive per hole, total charge 6,400 lbs fired instantaneously. At 720 ft distance the probable particle velocity is calculated to be 2.97 ips using Eq. 3.14.

Illustration $\begin{array}{cc} 00000 & 00000 \\ 00000 & 00000 \end{array}$

This particle velocity far exceeds safe limits, so two delays ms1 and ms2 were introduced to cut down the vibration level. This divided the shot into two smaller shots of 3,200 lbs per delay so that 3,200 lbs is the maximum vibration generating charge not the total 6,400 lbs. The 3,200 lb packages are separate and distinct because of the delays. The probable particle velocity is calculated to be 1.71 in. per sec.

Illustration $\begin{array}{cc} 00000 & 00000 \text{ ms2} \\ 00000 & 00000 \text{ ms1} \end{array}$

Consider the introduction of two more delays, ms3 and ms4. This divided the shot into four smaller shots of 1,600 lbs per delay. The probable particle velocity is calculated to be 0.98 in. per sec.

Illustration   ms3 00000   00000 ms4
                 ms1 00000   00000 ms2

Conditions are now safe. Delays have reduced the vibration and eliminated the hazards and potential for damage.

Why does delay blasting work? How does it reduce vibration? To understand this, one must first distinguish between particle velocity and propagation velocity.

Propagation velocity versus particle velocity. Propagation velocity is: how fast the seismic wave travels through the earth from the shot where it originated to the sensor and beyond. Values for compressional wave propagation velocity range from 1,000 to 20,000 ft per sec. For a given area, the value is approximately constant.

Particle velocity is radically different. As a seismic wave passes a point, the rock particle vibrates while staying in place. A rock particle at point A does not move away from A, it oscillates around A in an elliptical orbit and returns to point A when it stops vibrating. The orbital motion is only a few thousandths of an inch and its velocity is in inches per second. A simple example of particle velocity and *propagation* velocity is the motion of a fisherman in a boat. A passing speed boat generates a wave that passes under the fisherman, causing his boat to oscillate up and down. The up and down oscillation of the boat is the particle motion. The speed at which it oscillates is analogous to particle velocity. Particle velocity is what the seismograph measures. Values for particle velocity generated by blasting fall in inches per second (ips) range, usually fractions of an inch per second, e.g., 0.48 ips.

Vibration occurs when a seismic wave travels through the ground. Delay blasting works because it reduces the ground vibration at any instant of time. If the total charge is fired, one large wave train is generated. If delays are used, a series of small wave trains are generated, one for each delay. Figure 3.10 shows the seismic waves from the total charge and from delayed charges. Because it separates the seismic waves, each delay generates its own wave. The wave generated by the first delay has traveled a considerable distance because of its propagation velocity before the second delay has fired. The second seismic wave travels at the same propagation velocity as the first so it can never catch up to the first wave. Precise timing for the firing of each hole is essential for delay blasting to be effective. Figure 3.11 illustrates the process.

A necessary condition for successful delay blasting is that the seismic wave from any detonating blast hole shall clear all other blast holes before any of them are fired.

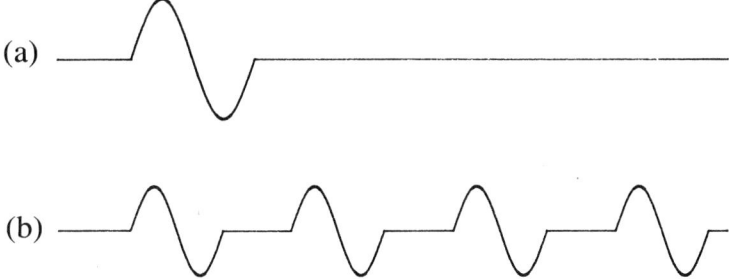

Figure 3.10 Simulated blast vibration pattern. (a) One large instantaneous shot; (b) same charge fired with four delays.

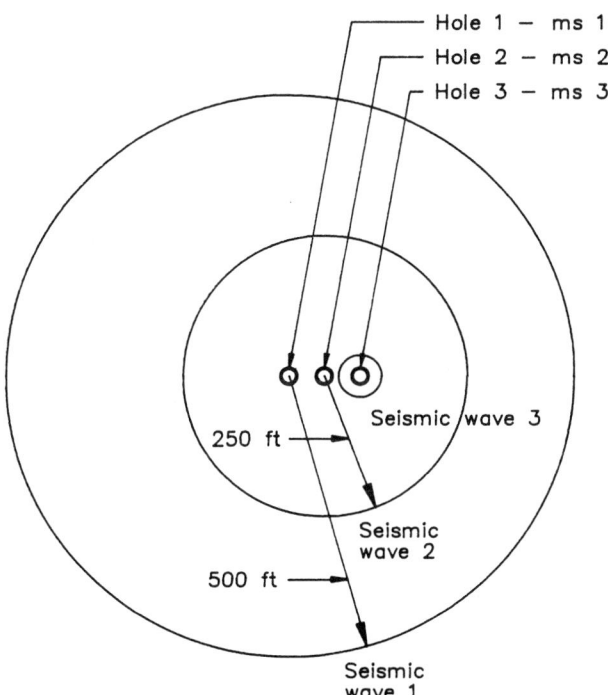

Figure 3.11 Seismic waves from delay blasting.

Lack of timing control and its effects.   Delay blasting is effective when the delays function properly. If they do not fire at the designated or nominal time specified on each cap, problems can arise. Consideration here will be limited to vibration effects such as increased motion (higher peak particle velocity) directional effects, variations in frequency, and wave length.

Consider two holes: hole 1 and hole 2, *not necessarily adjacent.* Hole 1 is fired and generates a seismic wave that travels outward from the hole in a circular pattern on the surface of the ground. Hole 2 fires before the wave from hole 1 arrives and generates its circular waves. The two waves meet at a point between the two holes and the resultant motion is the sum of the two motions and the vibration level will be significantly increased. The effect is indicated in Fig. 3.12.

This increase in wave motion and peak particle velocity can be directional.

If the cap firing is simultaneous, so that the waves meet midway between the holes, the maximum motion will lie along a line perpendicular to the line joining the holes. Figure 3.13 illustrates this.

The maximum motion will occur at the midpoint between the holes. As the wave travels further successive points of intersection are indicated. Here, the resultant motion will be the sum of the individual motions but it will be less than at the midpoint maximum because the waves have traveled farther. The direction of this line of maximum motion is perpendicular to the line of centers of the holes.

Consider another case in which hole 2 fires at the instant that the seismic wave from hole 1 arrives. The energies and motions will again be added and the waves will coincide along the line of centers of the holes. Figure 3.14 illustrates this effect.

The maximum motion will occur at the coincidence of the two seismic waves which will lie along the projected line of centers of the holes. The motion will have its maximum at the instant at which hole 2 fires coinciding with the arrival of the seismic wave 1. At successive points of coincidence the motion will be smaller than this maximum because the waves have traveled farther.

A more general case lies between the two cases considered, maximum motion perpendicular to the line of centers and along the line of centers.

For this case hole 2 fires so that the seismic waves meet somewhere between the midpoint of the holes and coincidence at hole 2. This is illustrated in Figure 3.15.

This line of added motion is curved and is hyperbolic in character. Its direction depends on the time delay between the caps and covers all azimuths. The

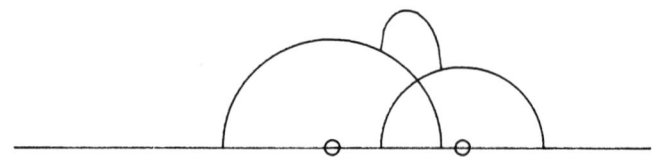

Figure 3.12   Increased amplitude due to interacting vibrations.

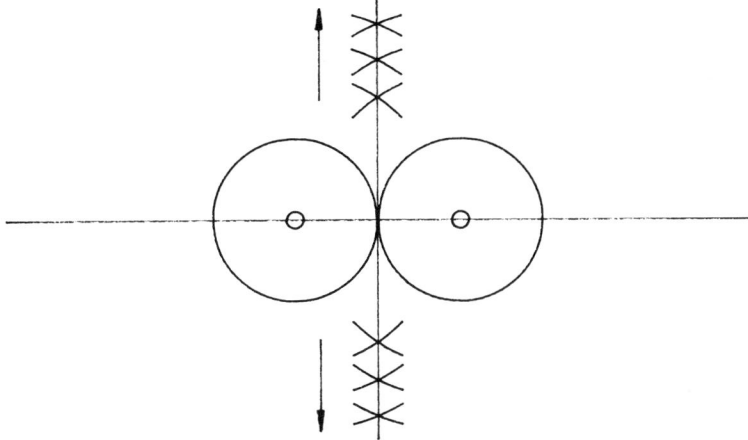

Figure 3.13 Vibration directionality perpendicular to the shot line.

curvature could be reversed if hole 2 fired early enough that seismic wave 2 had traveled past the midpoint before it reached seismic wave 1. Thus, the line of increased motion could lie anywhere.

Frequency and Wave Length Effects

When this line of increased motion occurs, what are its dimensions and how large an area is affected? It will cover a space of the order of one to two wave lengths. Wave length is defined as propagation velocity multiplied by the wave period.

$$L = VT \text{ (Eq. 3.3)}$$

where

L = wave length in ft
V = propagation velocity in ft per sec
T = wave period in sec

For a wave of period 1/10 sec and propagation velocity 2,000 ft per sec the wave length is 200 ft.

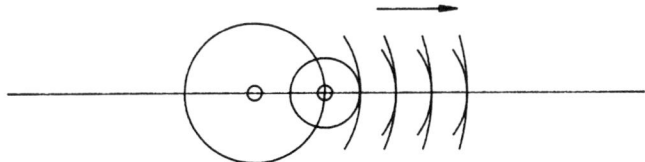

Figure 3.14 Vibration directionality along the shot line.

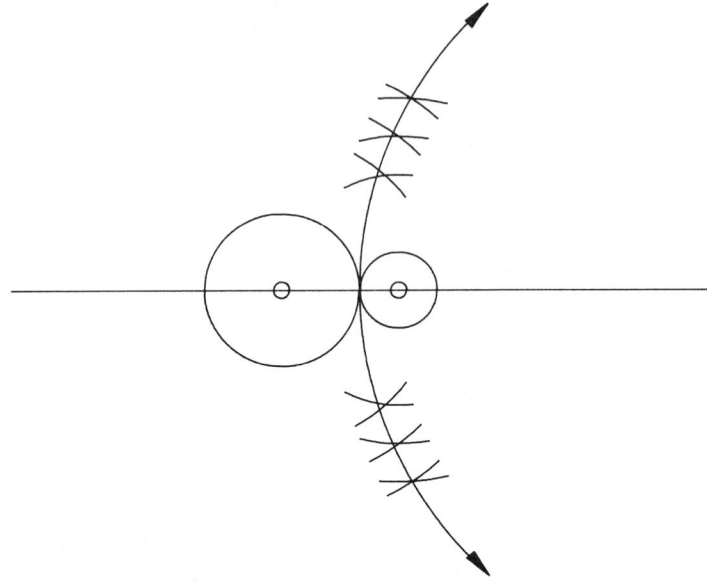

Figure 3.15 Vibration directionality, general case, covers all possible azimuths.

Assuming the waves are approximately the same (Fig. 3.16), at maximum coincidence the motion would be doubled but the wave length will be that of either wave since they are the same (Fig. 3.17).

When the waves first begin to meet, the composite form will appear similar to that shown in Fig. 3.18.

This form will be repeated after the maximum has occurred when the waves pass complete coincidence and begin to separate each into its own distinct form. Thus, there is a repetition or periodicity whose wave length approaches the sum of the two waves lengths or double that of the individual waves if they are the same. Also, the wave length of the composite motion varies from a single wave length to approximately double the single wave length. The converging and diverging wavelets are shown in Fig. 3.19 and the resulting composite motion is shown in Fig. 3.20.

The wave period and the frequency are each effected. At the point of maximum coincidence the period and frequency are those of the single wave. Since the period may approach double that of a single wave, the frequency will be cut approximately in half.

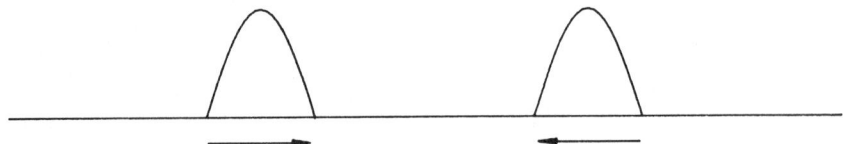

Figure 3.16 Converging equal wavelets.

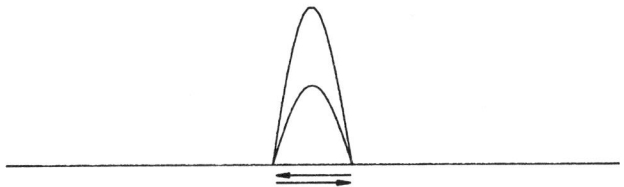

Figure 3.17   Composite wave motion at maximum coincidence.

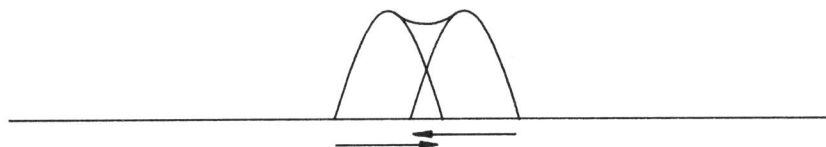

Figure 3.18   Converging wavelets.

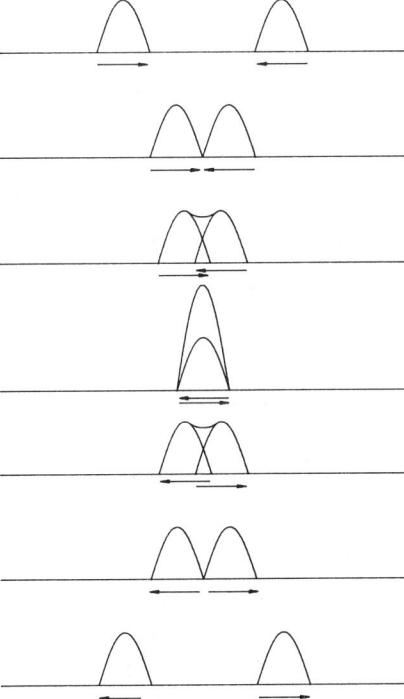

Figure 3.19   Converging and diverging wave interaction.

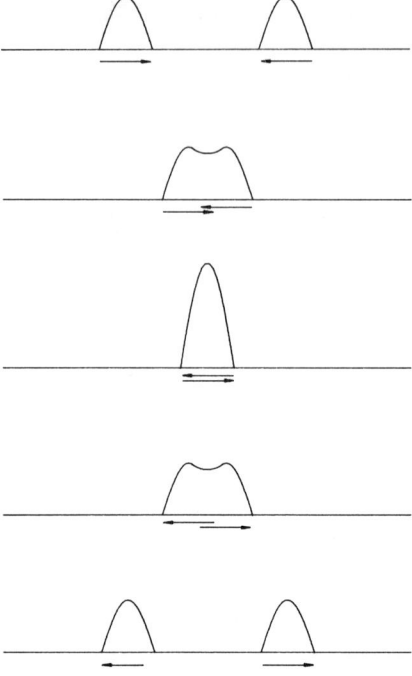

Figure 3.20   Composite motion.

The significant points here are that there can exist

1. a region of increased seismic motion and hence increased peak particle velocity with maximum at the center, minimum at the edges of the resultant combined waves
2. the region in which this occurs, the order of two wave lengths wide approximately 400 to 800 ft depending on propagation velocity and wave period
3. wave periods will be increased to approximately double with a corresponding lowering of the frequency to half
4. a region of high-seismic risk because of the increased motion and reduced frequency of vibration

Scaled distance.   The heart of the Propagation Law developed by the U.S. Bureau of Mines is $d/w^{0.5}$ which has been called scaled distance. It provides a practical and effective means to control vibration.

Scaled distance, written Ds, like ordinary distance, is safer when the value is large, more hazardous when it is small. Large values ($Ds > 50$) indicates safe vibration conditions—that is, low probability of damage—while small values ($Ds < 25$) indicates greater hazard with a higher probability of damage. The value

Ds = 50 originally proposed by the Bureau of Mines, was considered so safe that seismograph measurement was not necessary. It is a conservative limit, but many regulatory agencies are using a scaled distance of 60 or greater for increased safety.

Using the modified propagation law, the probable particle velocities can be calculated for these scaled distance values Ds = 50 and Ds = 60.

$$V = 100\ (50)^{-1.6}$$
$$V = 100\ (60)^{-1.6}$$
$$V = 0.19\ \text{ips}$$
$$V = 0.14\ \text{ips}$$

Scaled distance is a simple calculation once the distance and charge weight are known. Its usefulness lies in that the operator can compare the value for Ds with the regulatory value, and make a judgment as to the relative safety of the vibration. Examples of this are given here.

$$\text{Regulatory } Ds = 60$$

Charge Weight	Distance	Ds	
9	200	66	safe
175	800	60.5	safe
1500	2100	54.2	not safe

Safe distances or safe charges can be calculated using scaled distance. Sample calculations are illustrated here.

$$\text{Distance calculation for } \frac{D}{W^{0.5}} = 50$$

$$D = 50\ W^{0.5}$$

W lb	D ft
25	250
169	650
900	1500

$$\text{Charge calculation for } \frac{D}{W^{0.5}} = 50$$

$$W = (D/50)^2$$

D ft	W lb
50	1
200	16
900	324

Consider an example: assume a regulatory statute $Ds = 60$. A strip mine normally uses a charge of 1,600 lb per delay. A new housing development is starting at a distance of 2,200 ft. Is the strip mine in compliance?

$$Ds = \frac{D}{W^{0.5}}$$

$$60 = \frac{2200}{1600^{0.5}}$$

$$60 > 55 \text{ non compliance}$$

What charge will bring the strip mine into compliance?

$$60 = \frac{D}{W^{0.5}}$$

$$60 = \frac{2200}{W^{0.5}}$$

$$W = 1344 \text{ lb}$$

Any charge weight per delay of 1,344 lbs or less will be in compliance.

The strip mine is considering asking for a variance on the regulation since it cannot shoot effectively with less that 1,600 lbs per delay. The variance is denied. What then is the minimum distance for compliance?

$$60 = \frac{d}{(1600)^{0.5}}$$

$$d = 2400$$

The strip mine then is asking for a setback of 200 ft from the 2,200 foot property line so that no structure will be closer than 2,400 ft to the blasting. If agreed the strip mine may have to pay for the additional 200 ft width of property which cannot be mined and is only a buffer zone.

PROBLEMS

1. The period of a seismic wave is 0.125 sec. What is the wave frequency?
2. The above wave with period 0.125 sec has a propagation velocity of 4000 ft per sec. What is the wave length?
3. The maximum vibration from a blast is registered on the L-component with a particle velocity of 0.60 ips. At this same instant the V-component is 0.35 ips and the T-component is 0.22 ips. What is the vector sum particle velocity at this instant?
4. What does a seismic transducer do?
5. In sinusoidal motion the velocity is given by the equation

$$V = 2\pi f A \cos (2\pi ft)$$

(a) if $A = 0.025$ in.
    $f = 8$ Hz

Find the maximum peak particle velocity.

  (b) Find the maximum particle velocity if
  $A = 0.0012$ in.
  $f = 50$ Hz

6. Why is the maximum single amplitude used in vibration analysis instead of peak to peak or double amplitude?

7. If the peak particle velocity of a blast is $V = 0.82$ ips and the frequency is 12 Hz
  (a) what is the displacement A
  (b) what is the acceleration in in. per sec^2
  (c) what is the acceleration in g's?

8. On a quarry shot, the hole is loaded with 430 lbs of explosive per delay. The nearest structure is a home 825 ft from the shot.
  (a) Calculate the expected particle velocity using

$$V = 100 \ (d/w^{0.5})^{-1.6}$$

  b. If similar shots are fired and since the particle velocity may vary ±60%, what is the maximum particle velocity that could occur?

9. A strip mine is blasting based on a scaled distance of 32. What is the probable particle velocity generated by the blast?

10. Calculate the scaled distance values $D/W^{0.5}$ for the following explosive charges and distance:

d ft	w lb	Ds
180	25	
625	195	
1490	1020	

11. Using a scaled distance of 50, calculate the safe distance for the following explosive charges:

w lb	d ft
6.25	
84	
325	

12. Using a scaled distance of 50, calculate the permissible safe charge for the following distances:

d ft	w lb
28	
154	
638	

13. Many blast regulations specify a scaled distance value of 50 or 60. Calculate the expected particle velocity corresponding to these two values using the Eq. $V = 100 (Ds)^{-1.6}$

14. Using the blast seismogram, reproduced in Fig.3.4 measure the peak particle velocity on each component using a calibration value of one division equals one in/sec.

REFERENCES

"Blast Regression Analysis," Computer Software, Precision Blasting Systems, Montville, OH, 1987.

BOLLINGER, G.A., "Blast Vibration Analysis," Carbondale, IL: Southern Illinois University Press, 1971.

BRUEL and KJAER, "Acoustic Noise Measurements," 1979.

DUPONT, E.I. de Nemours and Co. Blasters Handbook 1977, *Blasters Handbook* (175th Anniversary Ed.), Wilmington, DE: Author.

DUVALL, W.I., "Design Criteria for Portable Seismographs," RI 5708, U.S. Bureau of Mines, 1961.

General Radio "Handbook of Noise Measurement," 7th ed., 1973.

KONYA, C.J. and WALTER, E.J., "Blasthole Timing Controls Vibration, Airblast and Flyrock," *Coal Mining,* January 1988.

KONYA, C.J. and WALTER, E.J., "Timing Controls Blasting Effects," *Rock Products,* June 1988.

LEET, L.D., "Earth Waves," Harvard University, 7th ed., Cambridge, MA, 1950.

NICHOLLS, H.R., JOHNSON, C.F. and DUVALL, W.I., "Blasting Vibrations and Their Effects on Structures," Bulletin No. 656, Washington, DC: U.S. Bureau of Mines, 1971.

4

Commercial Explosives Products

CHARACTERISTICS OF EXPLOSIVES

There are many different types of explosives used in commercial blasting. They can be classified as dynamites, slurry and ANFO. The selection of an explosive is based on three criteria: its ability to function properly in the proposed environment, the performance characteristics of the explosive, and cost. Our selection criteria is based on obtaining the explosive which will function under the specific conditions of the blast at the lowest possible cost. Explosives differ in the following ways:

Minimum diameter in which detonation will occur

The ability to resist water and water pressure

Generation of toxic fumes

Flammability

Ability to function under different temperature conditions

Input energy needed to start the reaction

Reaction velocity

Detonation pressure

Bulk density

Strength

Ability to remain in original configuration

If you wanted to purchase an explosive that had the best performance ratings in all of the above categories, it would also be the most expensive. On many jobs it is not necessary that explosives be the best in all the above categories. For instance, if we had a project with dry blast holes, we do not need a product that has a high-water resistance or if we are using large diameter blastholes, it is not important whether the explosive functions reliably in very small diameter blastholes. By selecting explosives that meet the particular job requirements, we can sacrifice some of the unneeded characteristics and purchase explosives that will function properly for the blast, at the lowest possible cost.

In our explosives selection process, we mentally evaluate each of the characteristics to determine which are important to us and which are not. On that basis, the explosive selection is made.

ENVIRONMENTAL CHARACTERISTICS OF EXPLOSIVES

The explosive must be able to function safely and reliably under the local environmental conditions. Six characteristics are considered in the selection of explosives which concern environmental factors: sensitiveness, water resistance, water pressure tolerance, fumes, flammability, and temperature resistance.

Sensitiveness. *Sensitiveness* is the characteristic of an explosive which defines its ability to propagate a stable detonation through the entire length of the charge and controls the minimum diameter for practical use. Sensitiveness is measured by determining the explosive's critical diameter. The term *critical diameter* is commonly used in the industry to define the minimum diameter of explosive column which will detonate reliably. All explosive compounds have a critical diameter. For some explosive compounds, the critical diameter may be measured in thousandths of an inch. On the other hand, other compounds may have critical diameters measured in inches. The diameter of the proposed borehole on a particular job will determine the maximum diameter of explosive column. The explosive diameter must be greater than the critical diameter of the explosive to be used in that borehole. Therefore, by preselecting certain borehole sizes, you may also eliminate certain explosive products from use on that particular job.

Table 4.1 is a table of sensitiveness for dynamites, slurry, and ANFO. Dynamites are broken into two separate categories: granular dynamite and gelatin dynamite. The slurries are also broken into two categories: cartridged and bulk loaded slurry which is pumped into the blastholes from large tanks on a bulk truck. Four types of ANFO are discussed in the table. Air-emplaced ANFO is primarily used in underground operations. It consists of bulk ANFO pneumatically loaded into blastholes. Pneumatic loading causes some particles of ANFO to break and, therefore, the load is densified. ANFO can also be poured or bulk loaded into dry blastholes. In wet hole applications, cartridged ANFO can be used. Heavy ANFO is a term used to describe a mixture of ANFO and slurry. The proportions of ingredients in the mixture can vary to accommodate energy requirements and environmental conditions.

TABLE 4.1 Sensitiveness (Critical Diameter)

Type	Critical Diameter		
	< 1 in.	1–2 in.	> 2 in.
Granular dynamite	x		
Gelatin dynamite	x		
Cartridged slurry	x	x	x
Bulk slurry		x	x
Air-emplaced ANFO	x		
Poured ANFO		x	
Packaged ANFO		x	x
Heavy ANFO			x

Water resistance. *Water resistance* is the ability of an explosive to withstand exposure to water without suffering detrimental effects in performance. Explosive products have two types of water resistance: internal and external. Internal water resistance is defined as water resistance provided by the explosive composition itself. As an example, some slurries (emulsions and water gels) can be pumped directly into boreholes filled with water. These explosives displace the water upward, but are not penetrated by the water and show no detrimental effects if fired within a reasonable period of time. External water resistance is provided not by the explosive materials itself, but by the packaging or cartridging into which the material is placed. As an example, ANFO has no internal water resistance; however, if it is placed in a sleeve or in a cartridge within a borehole, it can be kept dry and will perform satisfactorily. The sleeve on a cartridge provides the external water resistance for this particular product.

Water can dissolve or leach out some of the explosive ingredients, or it can cool the reaction to such a degree that the ideal products of detonation will not form even though the explosive is oxygen balanced. The emission of rust colored or yellow fumes from a blast indicates an inefficient detonation reaction and is frequently caused by water deterioration of the explosive. This condition can be remedied if a more water resistant explosive is used.

Manufacturers can describe the water resistance of a product in different ways. One way would be using terms such as excellent, good, fair, or poor (Table 4.2). When water is encountered in blasting operations, the explosive with at least a fair water resistance rating should be selected, and this explosive should be detonated as soon as possible after loading. If the explosive will be in water for an appreciable amount of time, it is advisable to select an explosive with at least a good water resistance rating. If water conditions are severe and the exposure time is significant, the prudent blaster may select an explosive with an excellent water resistance rating. Explosives with a poor water resistance rating should not be used in wet blastholes.

TABLE 4.2 Water Resistance

Type	Resistance
Granular dynamite	Poor to good
Gelatin dynamite	Good to excellent
Cartridged slurry	Very good
Bulk slurry	Very good
Air-emplaced ANFO	Poor
Poured ANFO	Poor
Packaged ANFO	Very good*
Heavy ANFO	Poor to very good

*Becomes poor if package is broken.

Water resistance ratings may also been given class ratings. For example:

Class	Time Maximum/Underwater (hours)
1	Indefinite
2	71
3	31
4	15
5	7
6	3
7	Less than 1

The descriptive method of rating water resistance is the one commonly seen on explosive data sheets. In general, product price is related to water resistance. The more water resistant the higher the product cost.

Water pressure tolerance.   The ability to remain unaffected by high static pressures is defined as *water pressure tolerance*. Some explosive compounds are densified and desensitized by hydrostatic pressures which result in deep boreholes. Combination of factors such as cold weather and small primers will contribute to failure. Under these conditions, energy release may be minimal with moderate static pressures which can occur on surface blasting operations. Problems with water pressure tolerance most often occur with slurry and heavy ANFO.

Fumes.   The *fume class* of an explosive is a measure of the amount of toxic gases produced in the detonation process. Carbon monoxide and oxides of nitrogen are the primary gases that are considered in the fume class ratings. Most commercial blasting explosives are oxygen balanced to minimize fumes and optimize energy release. However, fumes will still occur as a result of environmental conditions and the blaster should be aware of their existence. In underground applications, the problems that result from fumes especially with inadequate ventilation is obvious. Fumes can be hazardous to personnel working in deep cuts or trenches on surface blasting operations. Some conditions that can cause toxic fume production with oxygen balanced explosives are: insufficient charge diameter, inadequate water resistance, inadequate priming, and premature loss of confinement.

Gases such as carbon monoxide and nitrogen oxides are the most common fumes produced.

The Institute of Makers of Explosives (IME) have adopted a method of rating fumes and the test is conducted by the Bichel Gauge method. The cubic feet of poisonous gases released per 200 grams of explosives is measured. If less than .16 cu ft of toxic fumes are produced per 200 grams of explosive, the fume class rating would be 1. If .16 to .33 cu ft of poisonous gases are produced, the fume class rating is 2, and if .33 to .67 cu ft of poisonous gases are produced, the fume class rating is 3. Typical products are qualitatively rated in Table 4.3.

Strictly speaking, carbon dioxide is not a fume since it is not a toxic gas. Deaths have occurred, however, because of the generation of large amounts of carbon dioxide during blasting in confined areas. Although carbon dioxide is not poisonous, it is produced in large quantities in most blasts and it has the effect of causing the involuntary muscles of the body to stop working. In other words, the heart and lungs would stop working in high concentrations of carbon dioxide. If concentrations are 18% or higher in volume, death can occur by suffocation. An additional problem with carbon dioxide is that it has a specific gravity of 1.53 as compared to air; therefore, it remains in low places in the excavation and stays in that location for a long period of time when there is little air movement.

Flammability.   The *flammability* of an explosive is defined as the characteristic that deals with the ease of initiation from spark, fire, or flame. Some explosive compounds will explode from a spark while others can be burned and will not detonate. Flammability is an important consideration for safety in storage, transportation and use of explosives. Some explosives, although very economical,

TABLE 4.3 Fume Quality

Type	Quality
Granular dynamite	Poor to good
Gelatin dynamite	Fair to very good
Cartridged slurry	Good to very good
Bulk slurry	Fair to very good
Air-emplaced ANFO	Good*
Poured ANFO	Good*
Packaged ANFO	Good to very good
Heavy ANFO	Good*

*Can be poor under adverse conditions

have lost their marketability because of flammability. A good example is LOX, liquid oxygen and carbon, which was used in the 1950s as a blasting agent. Its flammability and inherent safety problems caused its demise in the USA. Most explosive compounds used today are not as flammable as LOX; however, accidents still occur because of flammability.

Over the past two decades, explosive products, in general, have become less flammable. Some manufacturers indicate that certain products can be burned without detonation in quantities as large as 40,000 lbs. The problem results because many blasters are given a false sense of security. Some believe that all products today are relatively nonflammable. This false sense of security has led to the death of people who have been careless with explosives and have assumed that flammability is not a problem. All explosive compounds should be treated as highly flammable. There should not be smoking during the loading process, and if explosives are to be destroyed by burning, the guidelines produced by the IME should be followed regardless of the type of explosive involved.

Temperature resistance.   Explosive compounds can suffer in performance if stored under extremely hot or cold conditions (Table 4.4). Under hot storage conditions, above 90°F, many compounds will slowly decompose or change properties and shelf life will be decreased. Storage of ammonium nitrate blasting agents in temperatures above 90°F can result in cycling, which will effect the performance and safety of the product.

Cycling of ammonium nitrate.   The chemical formula for ammonium nitrate (AN) is NH_4NO_3 or more simply written $N_2H_4O_3$. For its weight, AN supplies more gas volume upon detonation than any other explosive. In pure form, ammonium nitrate is almost inert and is composed of 60% oxygen by weight, 33% nitrogen, and 7% hydrogen. With the addition of fuel oil, the ideal oxygen balanced reactions for NH_4NO_3 is

TABLE 4.4   Temperature Resistance

Type	Between 0°F–100°F
Granular dynamite	Good
Gelatin dynamite	Good
Cartridge slurry	Poor below 40°F
Bulk Slurry	Poor below 40°F
Air-emplaced ANFO	Poor above 90°F
Poured ANFO	Poor above 90°F
Packaged ANFO	Poor above 90°F
Heavy ANFO	Poor below 40°F*

*Dependent on type and amount of slurry used in the blend.

$$3N_2H_4O_3 + CH_2 \rightarrow 3N_2 + 7H_2O + CO_2$$

Two characteristics make this compound both unpredictable and dangerous. Ammonium nitrate is water soluble and if uncoated can quickly attract water from the humidity in the atmosphere and slowly dissolve itself. For this reason, the spherical particles called *prills*, have a protective coating of silica flour (SiO_2) which offers some water resistance. The second and most important characteristic is a phenomena called cycling. *Cycling* is the ability of a material to change its crystal form with temperature. Ammonium nitrate will have one of five crystal forms depending on temperature.

1. Above 257°F cubic crystals exist.
2. Above 184°F and below 257°F tetragonal crystals exist.
3. Above 90°F and below 184°F orthorhombic crystals exist.
4. Above 0°F and below 90°F pseudotetragonal crystals exist.
5. Below 0°F tetragonal crystals exist.

The cycling phenomena can seriously effect both the storage and performance of any explosive which contains ammonium nitrate. Most dynamites, slurries, and ANFO, contain ammonium nitrate. Blasting agents and slurries are composed of large amounts of this compound. The two temperatures at which cycling will occur under normal conditions are 0°F and 90°F. Products which are stored over the winter or for a period of time during the summer will likely undergo some cycling. During the summer in a poorly ventilated powder magazine or storage bin located in the sun, the cycling temperature may be reached daily.

The effect of cycling on AN when isolated from the humidity in the air is that the prills break down into finer particles (Fig. 4.1). The prills under standard temperature conditions are made up of pseudotetragonal crystals. When the temperature exceeds 90°F, each crystal breaks into smaller crystals of orthorhombic structure. When the temperature again falls below 90°F, the small crystals break into even finer crystals of the pseudotetragonal form. This process can continue until the density is no longer near 0.8 g/cc, but can reach a density near 1.2 g/cc. The density increase can make the product more sensitive and it will contain more energy per unit volume. When the density is above 1.2 g/cc, ammonium nitrate will no longer detonate.

To further complicate the situation, some cartridged blasting agents or those stored in bins may not efficiently exclude humidity. After the amonium nitrate has undergone cycling, the water-resistant coating is broken and the water vapor in the air condenses on the particles. As cycling continues, water collects on the particles and the mass starts to dissolve. Recrystalizing into large crystals can occur when temperatures drop.

Ammonium nitrate may have very dense areas and areas of large crystals after

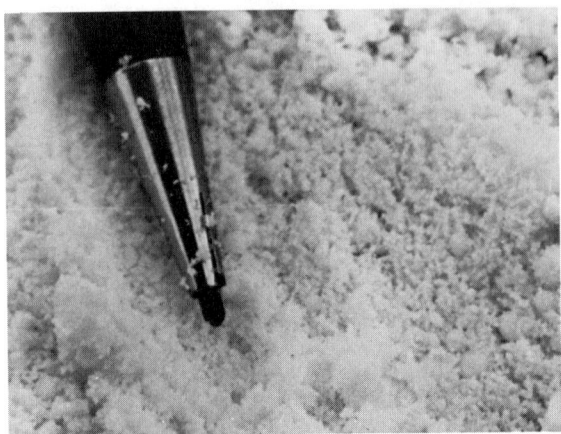

Figure 4.1 Picture of cycled AN. (Courtesy of IRECO Inc.)

cycling. The performance of this product may range from that of a very powerful explosive to one that deflagrates or one that will not shoot at all.

Cold resistance.   Cold conditions can also effect the performance of some explosive products. Most dynamites and ANFO blasting agents will not freeze under ordinary exposure to the normal temperatures encountered in the USA. This is because the manufacturers have added ingredients to these products that allow them to perform properly, in spite of the cold weather. Some products may stiffen and become firm after prolonged exposure to low temperatures and may become more difficult to use in the field.

Slurry explosives, which include water gel and emulsions, can have serious detonation problems if stored in cold temperatures and not allowed to warm up before they are detonated. Slurries are quite different from the other products previously mentioned, such as dynamite and blasting agents. Problems commonly occur because blasters had been accustomed to using explosives without concern for cold weather conditions. The slurry explosives do not all perform the same; some can be used immediately if stored at temperatures near 0°F. Others will not detonate if stored at temperatures below 40°F. The sensitivity of the product is affected. The priming procedure, which was employed when the product is at 70°F, may cause a misfire if the product is at 42°F. It is a good practice to consult the manufacturer's data sheet whenever any new product is introduced on the job. It is absolutely essential to consult that data sheet if any new slurry explosives are introduced, since their properties and performance with temperatures can vary greatly from one supplier to another. Warm-up charts (Fig. 4.2) are available to instruct the user. These charts provide information on how long a product should be left in a borehole before the blast is fired to give them sufficient time to warm up to temperatures at which detonation can occur.

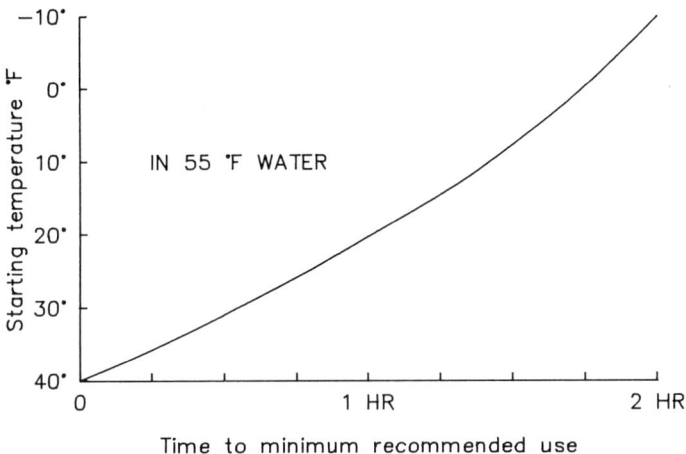

Figure 4.2   Warm-up chart.

PERFORMANCE CHARACTERISTICS OF EXPLOSIVES

Environmental conditions can eliminate certain types of explosives from consideration on a particular project. After the environmental conditions have been considered, you must also consider the performance characteristics of explosives. Characteristics of primary concern are sensitivity, velocity, detonation pressure, density, strength, and cohesiveness.

Sensitivity.   The *sensitivity* of an explosive product is defined by the amount of input energy necessary to cause the product to detonate reliably. This is sometimes called the minimum booster rating, or minimum priming requirements. Some explosives require little energy to detonate reliably. The standard Number 8 blasting cap will detonate dynamite and some of the cap sensitive slurry explosives. On the other hand, a blasting cap alone will not reliably initiate bulk loaded ANFO and some slurry. For these products, you would have to use a booster or primer in conjunction with the blasting cap to get reliable detonation.

Many factors can influence the sensitivity of a product. As an example, the sensitivity can be reduced by the effect of water in the blasthole, inadequate charge diameter, and temperature extremes. Sensitivity of a product defines its priming requirements, the primer size, and energy output. If reliable detonation of the main charge does not occur, fumes increase, ground vibration levels rise, blastholes geyser and flyrock can be thrown. Sensitivity is also a measure of the explosive's ability to propagate from cartridge-to-cartridge. It can be expressed as the maximum separation distance (in inches) between a primed donor cartridge and an unprimed receptor cartridge, where reliable detonation transfer will occur. Hazard sensitivity defines an explosive's response to the accidental addition of energy, such as fire or bullet impact (Table 4.5).

Velocity.   The *detonation velocity* is the speed at which the detonation moves through the column of explosive. It ranges from approximately 5,000 to 25,000 ft per sec for commercially used products. Detonation velocity is an important consideration for applications outside of a borehole, such as plaster shooting, mud capping, or shearing structural members. Detonation velocity, itself, has significantly less importance if the explosives are used in the borehole.

Detonation velocity can be used as a tool to determine the efficiency of the explosive reaction in field use. If a question arises as to performance of an explosive compound during actual field use, velocity probes can be inserted in the product. When the product is detonated, the reaction rate of the product can be measured and its performance judged by the recorded velocity. If the product is reacting at a velocity significantly lower than its rated velocity, it is an indication that its performance is not up to standard expectations. Typical explosive detonation velocities are given in Table 4.6.

Detonation pressure.   The *detonation pressure* is the near instantaneous pressure derived from the shock wave moving through the explosive compound (Table 4.7). When initiating one explosive with another, the shock pressure from the primary explosive is used to cause initiation in the secondary explosive. Detonation pressure can be related to borehole pressure but it is not necessarily a linear relationship. An explosive with similar detonation pressures will not necessarily have equal borehole pressure or gas pressure. Detonation pressure is calculated mathematically.

The detonation pressure is related to the density of the explosive and its reaction velocity. When selecting explosives for primers, detonation pressure is an important consideration. Methods to approximate detonation pressure and their relationship to priming will be discussed in Chap. 6, Primer and Booster Selection.

TABLE 4.5   Sensitivity

Type	Hazard Sensitivity	Performance Sensitivity
Granular dynamite	Moderate to high	Excellent
Gelatin dynamite	Moderate	Excellent
Cartridged slurry	Low	Good to very good
Bulk slurry	Low	Good to very good
Air-emplaced ANFO	Low	Poor to good*
Poured ANFO	Low	Poor to good*
Packaged ANFO	Low	Good to very good
Heavy ANFO	Low	Poor to good*

*Heavily dependent on field condition.

TABLE 4.6 Detonation Velocity (fps)

Type	Diameter		
	1 ¼ in.	3 in.	9 in.
Granular dynamite	7–19,000		
Gelatin dynamite	12–25,000		
Cartridged slurry	13–19,000	14–19,000	
Bulk slurry		14–19,000	12–19,000
Air-emplaced ANFO	7–10,000	12–13,000	14–15,000
Poured ANFO	6–7,000	10–11,000	14–15,000
Packaged ANFO		10–12,000	14–15,000
Heavy ANFO			11–19,000

Density.   The *density* of explosive is important because explosives are purchased, stored and used on a weight basis. Density is normally expressed in terms of specific gravity, which is the ratio of explosive weight to water weight for the same volume. The density of an explosive determines the weight of explosive that can be loaded into a specific borehole diameter. The difference in energy on a unit weight basis is nowhere near as great as the difference in energy on a volume basis. When hard rock is encountered and drilling is expensive, a denser product of higher cost is often justified. Typical specific gravity values for explosive products are given in Table 4.8.

The specific gravity of the explosive is commonly used as a tool to approximate strength and design parameters between explosives of different manufacturers and different generic families. In general terms, the higher the explosive density, the more energetic the product. The specific gravity of commercial products range from about 0.8 to 1.6.

Another useful expression of density is what is commonly called *loading density* or the weight of explosive per linear foot of charge in a specified diameter. Loading density is used to determine the total pounds of explosive which will be used per borehole and per blast.

TABLE 4.7 Detonation Pressure

Type	Detonation Pressure (kbars)
Granular dynamite	20–70
Gelatin dynamite	70–140
Cartridged slurry	20–100
Bulk slurry	20–100
Poured ANFO	7–45
Packaged ANFO	20–60
Heavy ANFO	20–90

TABLE 4.8 Density (g/cc)

Type	Density (g/cc)
Granular dynamite	0.8–1.4
Gelatin dynamite	1.0–1.6
Cartridged slurry	1.1–1.3
Bulk slurry	1.1–1.6
Air-emplaced ANFO	0.8–1.0
Poured ANFO	0.8–0.85
Packaged ANFO	1.1–1.2
Heavy ANFO	1.1–1.4

An easy method to calculate loading density is

$$de = 0.34 \ SGe \ De^2 \tag{4.1}$$

where

$$de = \text{loading density in lb/ft}$$
$$SGe = \text{specific gravity of explosive}$$
$$De = \text{diameter of explosive}$$

Let us determine the loading density of an explosive which has a charge diameter of 3 in. and a specific gravity of 1.2.

$$de = 0.34 \ SGe \ De^2$$
$$de = 0.34 \times 1.2 \times 3^2$$
$$de = 3.67 \ \text{lb/ft}$$

Strength.   The term strength refers to the energy content of an explosive which in turn is the measure of the force it can develop and its ability to do work. Strength has been rated by various manufacturers both on a equal weight and an equal volume basis, and are commonly called weight strength and cartridge or bulk strength. There is no standard strength measurement method universally used by explosives manufacturers. Instead many different strength measurement methods exist such as the ballistic mortar test, seismic execution values, strain pulse measurement, cratering, calculation of detonation pressures, calculation of borehole pressures, determination of heat release, and bubble energy tests. However, none of these methods can be used satisfactorily for blast design purposes. Theoretical or controlled test data does not necessarily represent the useful work energy delivered in the borehole. The delivered energy varies from one type of explosive to another and for the same explosive in different borehole diameters. For example in one diameter of boreholes a product may be 60% efficient, while in another diameter it may be 90% efficient. Strength ratings are misleading and don't accurately compare rock fragmentation effectiveness with explosive type. In general, one can say that strength ratings are only a tool used to identify the end results and associate them with a specific product.

One type of strength rating, the bubble energy test which determines the energy of the expanding gas is used by some for design purposes. The bubble energy test does produce some results which can be used for calculating blast design dimensions.

Cohesiveness.   *Cohesiveness* is defined as the ability of the explosive to maintain its original shape. There are times when the explosive must maintain its original shape and others when it should flow freely. For example, when blasting in cracked or broken ground, one definitely wants to use an explosive that will not flow into the cracked area causing holes to be overloaded. Conversely, in other applications such as in bulk loading, explosives should flow freely and not bridge the borehole nor form gaps in the explosive column.

EXPLOSIVES PRODUCTS

The products used as the main borehole charge can be broken into three generic categories: dynamite, slurries, and ANFO blasting agents (Fig. 4.3). A fourth, a very minor category will be added to the discussion which is the binary or two component explosives. Although the volume of binary explosives sold annually is insignificant when compared to the other major categories, its unique properties warrants its mention.

All the categories discussed in this section are high explosives because they will all detonate. On the other hand, one commonly hears some of these high explosives called by other names such as blasting agents. The term blasting agent does not detract from an explosive's ability to detonate or function as a high explosive. Blasting agent are a classification for the storage and transportation of some high explosives. They are less sensitive to initiation and, therefore, can be stored and transported under different regulations than would normally be used for more sensitive high explosives. The term high explosive refers to any product used in blasting that reacts at a speed faster than the speed of sound in the explosive material. The reaction must also be accompanied by a shock wave for it to be considered a high explosive.

The subclass of high explosives, called *blasting agents*, are materials or mixture which consists of a fuel and an oxidizer. The finished product as mixed and packaged for shipping cannot be detonated by a Number 8 blasting cap in a specific test prescribed by the Bureau of Mines. Normally, blasting agents do not contain ingredients which, in themselves, are high explosives. Some slurries containing TNT, smokeless powder or other high-explosive ingredients may be classed as a blasting agent if they are insensitive to initiation by a Number 8 blasting cap.

Dynamite

Most dynamites are nitroglycerin based products. A few manufacturers of dynamite have products in which they substituted nonheadache producing high

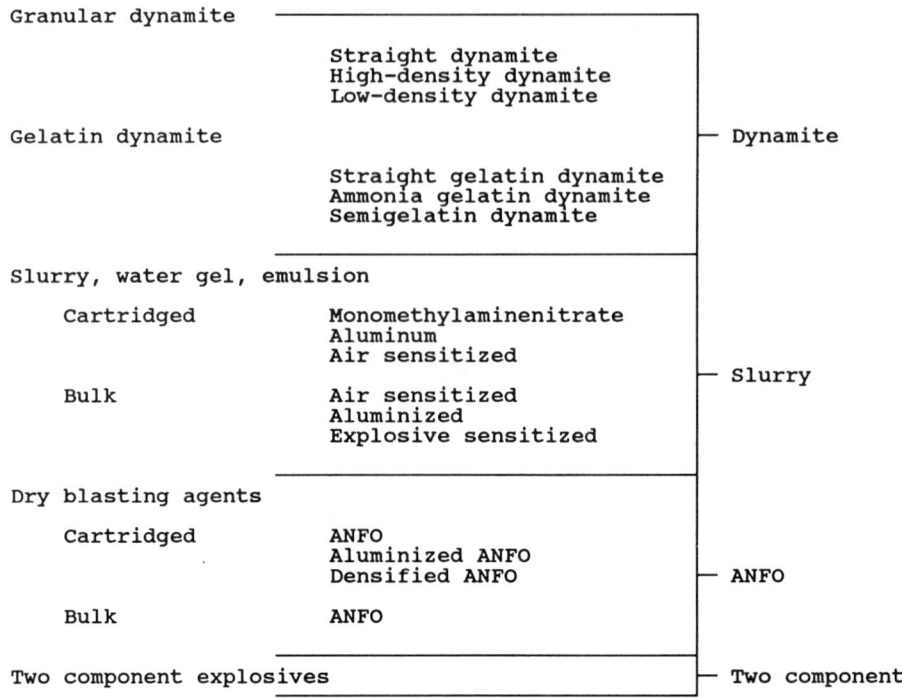

Figure 4.3 Types of explosives.

explosives such as nitrostarch or other nitro compounds for the nitroglycerin. Dynamites are the most sensitive of all the explosives used today. Because of the sensitivity, they offer an extra margin of dependability in the blasthole, since gaps in loading within the explosive column and many other environmental factors which cause other explosives to malfunction would not affect dynamite. Dynamite is somewhat more susceptible to accidental initiation because of its sensitivity.

Nitroglycerin was the first high explosive used in commercial blasting. It has a specific gravity of 1.6 and detonation velocity of approximately 25,000 ft per sec. Nitroglycerin is extremely sensitive to shock, friction, and heat, which makes its use in liquid form extremely hazardous. In Sweden in 1867, Nobel found that if this hazardous liquid was absorbed into an inert material, the resulting product would be safe to handle and would be much less sensitive to shock, friction, and heat. This product was called dynamite.

Within the dynamite family, there are two major subclassifications, granular dynamite and gelatin dynamite. Granular dynamite is a compound which uses nitroglycerin as the explosive base, whereas gelatin dynamite uses a mixture of nitroglycerin and nitrocellulose which produces a rubbery waterproof compound.

Granular dynamite.   Under the granular dynamites classification, there are three subclassifications which are straight dynamite, high-density extra dynamite and low-density extra dynamite (Fig. 4.4).

Straight dynamite.   Straight dynamite consists of nitroglycerin, sodium nitrate, carbonaceous fuels, sulfur and antacids. The term straight means that a dynamite contains no ammonium nitrate. Straight dynamite is the most sensitive commercial high explosive in use today. It should not be used for construction or mining applications since its sensitivity to shock could result in sympathetic detonation in wet blastholes rather than the initiation from the caps within the hole. On the other hand, straight dynamite is an extremely valuable product for blasting dirt ditches. Sympathetic detonation is an attribute in ditching because it eliminates the need for a detonator in each and every hole. In ditching applications, normally one detonator is used in the first hole and all other holes are fired by sympathetic detonation. Although ditching dynamite is more costly than other types of dynamite, it can save a considerable amount of money and time for ditching applications.

High-density extra dynamite.   This product is the most widely used dynamite. It is similar to straight dynamite except that some of the nitroglycerin is replaced with ammonium nitrate. The ammonia or extra dynamite is less sensitive to shock and friction than the straight dynamite. It has found broad use in quarries, underground mines, and construction.

Low-density extra dynamite.   Low-density extra dynamites are similar in composition to the high-density products except that more nitroglycerin is replaced with ammonia nitrate. Since the cartridge contains a large proportion of ammonium nitrate, its bulk or volume strength is relatively low. This product is useful in weak rock or where a deliberate attempt is made to limit the *energy density*, the energy per lineal foot of borehole.

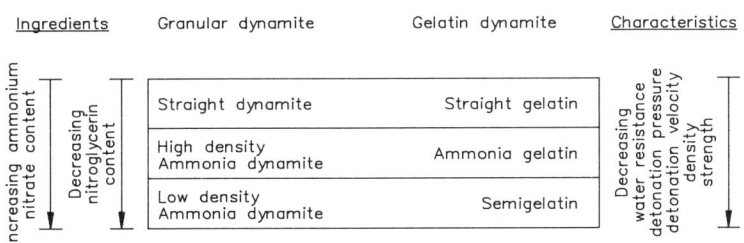

Figure 4.4   Characteristics of dynamite.

Gelatin dynamite. Gelatin dynamite used in commercial applications can be broken into three subclasses: straight gelatin, ammonia gelatin, and semigelatin dynamites.

Straight gelatin dynamite. Straight gelatins traditionally contain nitroglycerin and nitrocellulose with sodium nitrate and carbonaceous fuel and sometimes sulfur added. In strength, it is the gelatinous equivalent of straight dynamite. A straight blasting gelatin is the most powerful nitroglycerin-based explosive. A straight gelatin because of its composition would also be the most water resistant dynamite.

Ammonia gelatin dynamite. Ammonia gelatin is sometimes called special or extra gelatin. It is a mixture of straight gelatin with additional ammonium nitrate added to replace some of the nitroglycerin. Ammonia gelatin are suitable for wet conditions and are primarily used as bottom loads in small diameter blastholes. Ammonia gelatins do not have the water resistance of a straight gelatin. Ammonia gelatins are often used as primers for blasting agents.

Semigelatin dynamite. Semigelatin dynamites are similar in some respects to ammonia gelatins except that more of the nitroglycerin, nitrocellulose mixture, is replaced by ammonium nitrate. Semigelatin dynamites are less water resistant than the ammonium gelatins and more economical because of their lower costs. They do have more water resistance than the granular dynamites because of their gelatinous nature and are often used under wet conditions. They are sometimes used as primers for blasting agents.

Slurry Explosives

A *slurry explosive* is made of ammonium, calcium or sodium nitrate, a fuel sensitizer, which can either be a hydrocarbon or hydrocarbons with aluminum or in some cases explosive sensitizers such as TNT or nitrocellulose, along with varying amounts of water (Fig. 4.5). Slurries can be broken down into two categories: water gels and emulsions. Water gels require a sensitizer and a cross-linking agent while emulsions are composed of saturated solutions of ammonium nitrate and fuel in the form of an emulsion. An emulsion is somewhat different from a water gel in characteristics, but the composition contains similar ingredients and functions similarly in the blasthole (Fig. 4.6). In general, emulsions have a higher detonation velocity and in some cases, may tend to be wet or adhere to the blasthole causing difficulties in bulk loading. For discussion purposes, emulsions and water gels will be treated under the generic family of slurries.

Slurries, in general, contain large amounts of ammonium nitrate and are made water resistant through the use of gums, waxes, cross linking agents in water gels or emulsifying agents in emulsions. Numerous varieties of slurries exist, and it must be remembered that different slurries will exhibit different characteristics in the

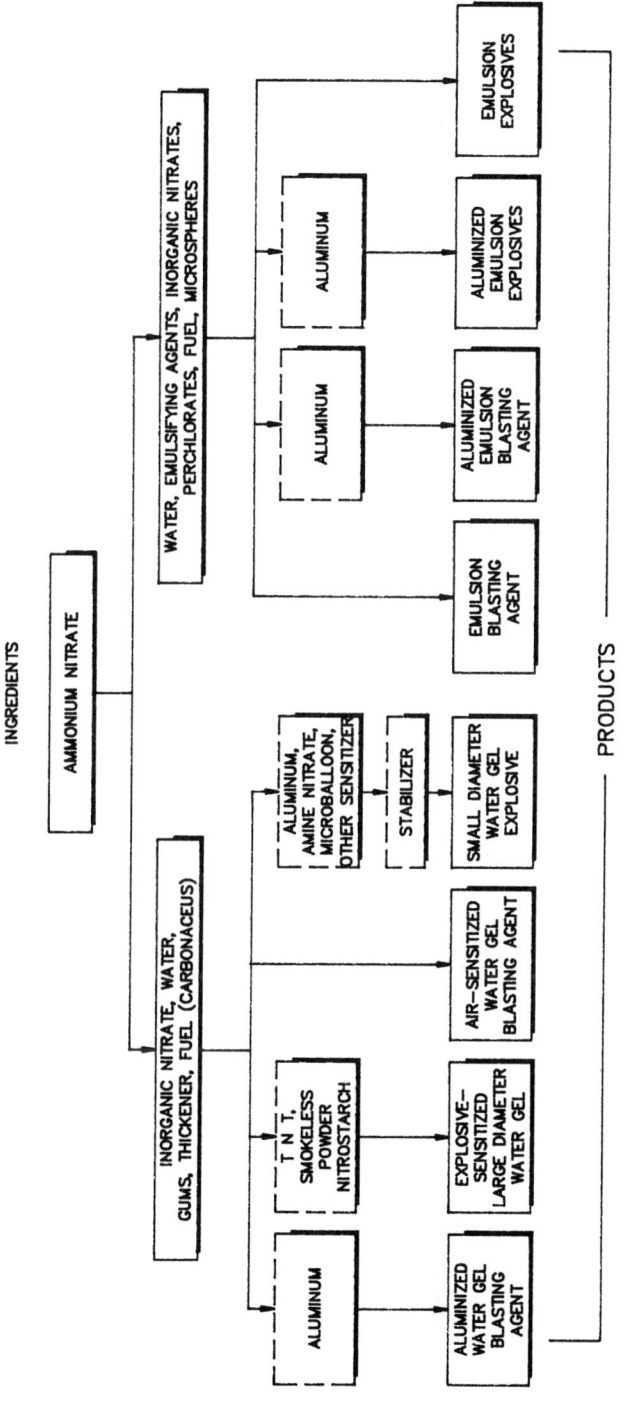

Figure 4.5 Composition of slurry explosives.

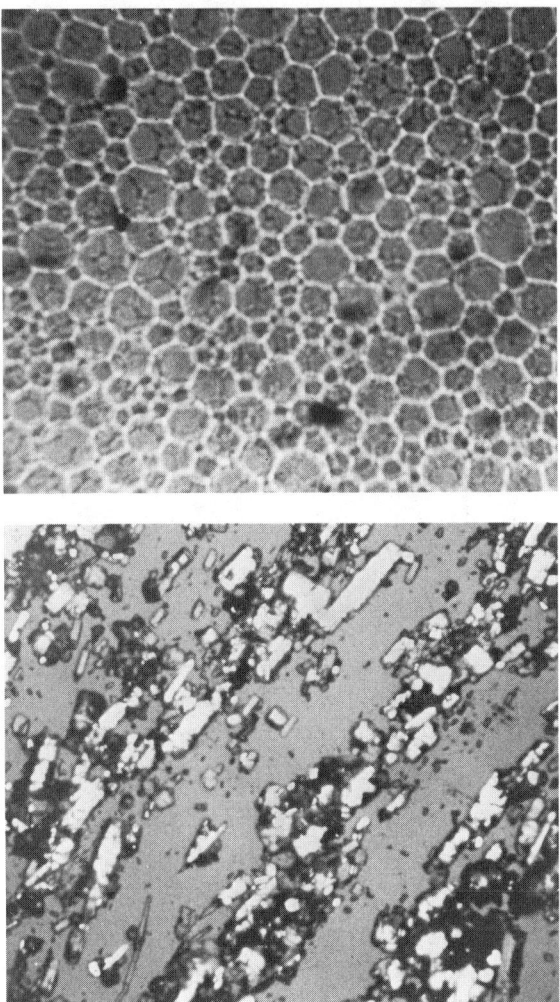

Figure 4.6 Emulsion (top) and water gel consistency (bottom). (Courtesy of Atlas Powder Company)

field. Some slurries may be classified as high explosives, while others are classified as blasting agents if they cannot be initiated by a Number 8 blasting cap. This difference in classification is important for magazine storage. An added advantage to slurries over dynamites is that they can be delivered as separate ingredients for on-site mixing. The separate ingredients brought to the job site in large tank trucks are generally nonexplosive until mixed at the blasthole. The bulk loading of slurries can greatly reduce the time and cost of loading large quantities of explosives (Fig. 4.7).

Slurries can be broken down into two general classifications: cartridge and bulk.

Figure 4.7   Bulk emulsion slurry. (Courtesy of Atlas Powder Company)

Cartridged slurries.   Cartridge slurries come in both large and small diameter cartridges. In general, cartridges less than 2 in. in diameter are normally made cap-sensitive so that they can be substituted for dynamite. The difference in temperature resistance of slurries and their lower sensitivity can cause problems when substituted for some dynamite applications. The blaster must be aware of these limitations before a one-for-one substitution is tried. The larger diameter cartridged slurries are not cap-sensitive and must be primed with cap-sensitive explosives. In general, large diameter slurries are the least sensitive. Cartridge water gels may be sensitized with monometholamine nitrate or aluminum, and air sensitized in the case of emulsions. Air sensitizing is accomplished by the addition of microspheres or entrapping air during the mixing process itself.

Bulk slurries.   Bulk slurries are sensitized by one of three methods. Air sensitizing can be accomplished by the addition of gassing agents which after being pumped into the blasthole produce small gas bubbles throughout the mixture. The addition of powdered or scrap grade aluminum, nitrocellulose or TNT will also increase sensitivity.

Slurries containing neither aluminum nor explosive sensitizers have the lowest cost. They are often the least dense and the least powerful. In wet conditions where mechanical dewatering is not practiced, low-cost slurries offer competition to packaged ANFO. Some low-cost slurries, however, can have less energy than ANFO. Aluminized slurries and those containing high-explosive sensitizers can develop more energy and are used for blasting harder dense rock. The alternative to using high-energy slurries is pumping blastholes, where possible, with submersible blasthole pumps (Fig. 4.8) and using polyethylene blasthole liners and ammonium nitrate as the explosive (Fig. 4.9). In many applications, the use of pumping with sleeves and ammonium nitrate will produce blasting costs which are significantly

Figure 4.8 Pumping blastholes.

less than would result from using the higher priced slurries. Pumping and sleeving supplies are available from explosive distributors.

Dry Blasting Agents

Dry blasting agents are the most common of all explosives used today. Approximately 80% of the explosives used in the USA are dry blasting agents. The term *dry blasting agent* describes any material in which no water is used in the formulation. Early dry blasting agents employed fuels of solid carbon or coal dust combined with ammonium nitrate in various forms. Solid fuels segregated during transportation and provided less than optimum blasting results. Diesel oil was substituted for the solid fuel. It was found that diesel oil mixed with porous ammonium nitrate prills gave the best overall blasting results. The term ANFO has become synonymous with dry blasting agents. An oxygen balanced mixture of ANFO is the

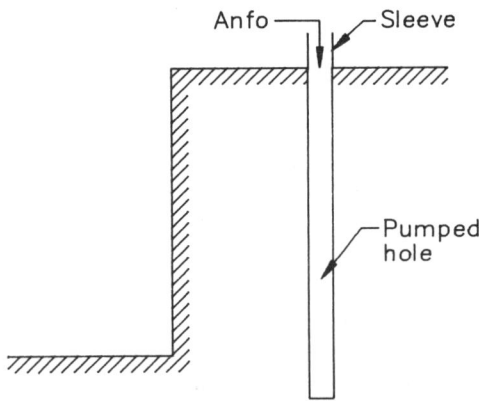

Figure 4.9 Sleeves for wet holes.

lowest cost source of explosive energy available today. Ground aluminum foil added to dry blasting agents increases the energy output and cost. Dry blasting agents can be broken down into two categories: cartridged and bulk (Fig. 4.10).

Cartridged blasting agents.   For wet-hole use, where blastholes are not pumped, an aluminized or densified ANFO cartridge can be used in place of bulk ANFO. Densified ANFO is made by either crushing approximately 20% of the prills and adding them back into the normal prill mixture or by adding inert metallic compounds to increase the bulk density of the cartridge. In both cases, the object is to produce an explosive with a specific gravity greater than 1.0 so that it will sink in water. Another type of ANFO cartridge is that which is made from the normal bulk ANFO with a density of 0.8. This cartridge will not sink in water; however, it is a cost advantage to use this type of cartridged ANFO when placing them in wet holes that were recently pumped and contain only small amounts of water.

Bulk ANFO.   *Bulk ANFO* is prilled ammonium nitrate with fuel oil added. It can be purchased in bags or bulk for bin storage. It is often either blown or augured into the blasthole from a bulk truck. The mixed ANFO can be placed in the truck for borehole loading or in some trucks the dry ammonium nitrate and diesel oil can be field-mixed as the material is being placed in the borehole. The blasting industry has a great dependence on dry-blasting agents because of the large volume

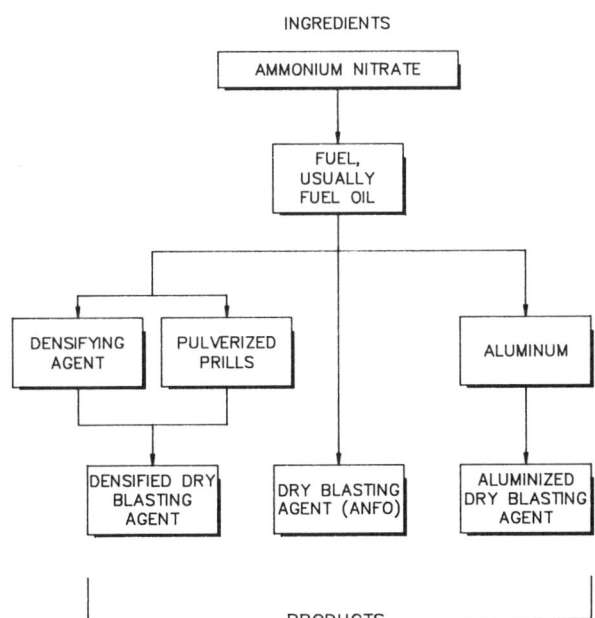

Figure 4.10 Composition of ANFO blasting agents.

used. Dry-blasting agents will not function properly if placed in wet holes for extended periods of time. For this reason, the blaster should know the limitations of his product.

Water resistance of ammonium nitrate.   Ammonium nitrate, which is bulk loaded into a blasthole, has no water resistance. If the product is placed in water and shot within a very short period of time, marginal detonation can occur with the production of rust colored fumes of nitrous oxide. The liberation of nitrous oxide is commonly seen on blasts involving bulk ammonium nitrate when operators have not taken the care to load the product in a proper manner which ensures that it will stay dry. Although a marginal detonation occurs, the energy produced is significantly less than the product would be capable of producing under normal conditions. We often see blastholes geysering, flyrock thrown, and other problems arising from using ammonium nitrate fuel oil mixtures in wet blastholes. If ammonium nitrate is placed in wet blastholes, ammonium nitrate will absorb water. When the water content reaches approximately 9%, it is questionable whether the ammonium nitrate will detonate regardless of the size primer used. Figure 4.11 indicates the effect of water content on the performance of ammonium nitrate. As water content increases, detonation velocity and explosive efficiency decreases.

Energy output of ANFO.   When ammonium nitrate fuel oil mixtures are made in the field, variations in oil content can easily occur. Bagged mixtures received from some distributors have similar problems. The amount of fuel oil

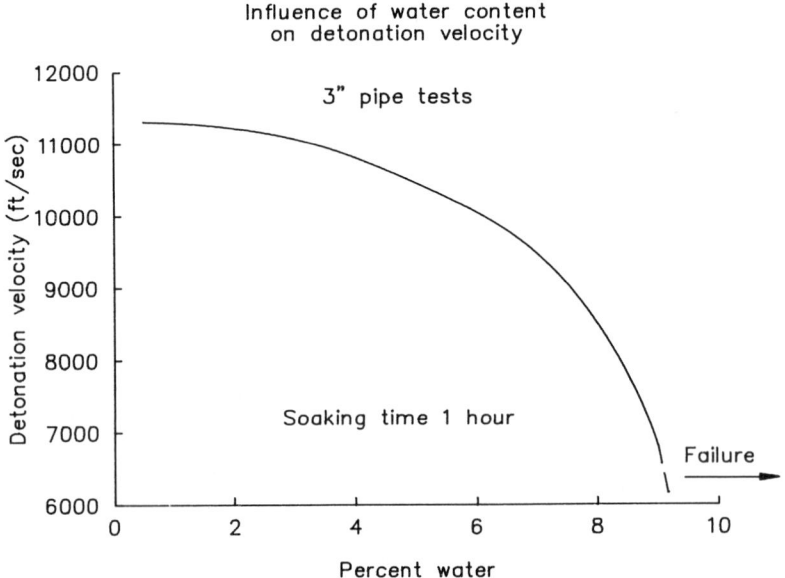

Figure 4.11 Influence of water on ANFO performance.

placed on the ammonium nitrate is extremely critical from the standpoint of efficient detonation (Fig. 4.12). To get the optimum energy release, you want about a 94% ammonium nitrate with a 6% diesel oil mixture. This would be approximately 3.3 quarts of fuel oil per 100 lbs of mixture. If for some reason, rather than the required 6% oil on the prills they contain only 2.0% oil, a significant amount of the energy is wasted and the explosive will not perform properly. Too little fuel will promote the formation of rust colored nitrous oxide fumes in dry holes. An excess of fuel oil is detrimental to the maximum energy output in ammonium nitrate fuel oil mixtures. It is, however, less detrimental than too little fuel. Figure 4.12 indicates the effect on theoretical energy at different fuel oil percentages. ANFO is more sensitive to initiation when under-fueled than when properly fueled.

Properties of blasting prills. Ammonium nitrate used for bulk loading comes in the form of prills. The prills are spherical particles of ammonium nitrate manufactured in a prilling tower with a similar process to that used in making bird shot for shot shells (Fig. 4.13). Ammonium nitrate prills are also used in the fertilizer industry. During times of explosive shortages, the blaster has often gone to feed mills and purchased fertilizer grade ammonium nitrate prills. There are differences between the fertilizer grade and the blasting grade prills. The blasting prill is considered a porous prill, which better distributes the fuel oil and results in much better performance on the blasting job. Table 4.9 indicates the difference in characteristics of fertilizer and blasting prills.

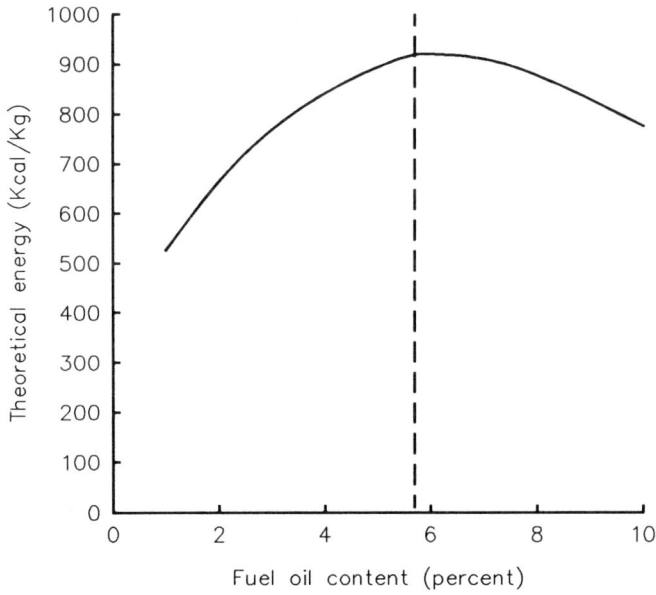

Figure 4.12 Theoretical energy output of ANFO at different fuel oil contents.

Figure 4.13 ANFO prills.

Heavy ANFO. *Heavy ANFO* or ammonium nitrate blends are mixtures of ammonium nitrate prills, fuel oil, and slurries. The advantage to heavy ANFO blends is that they can be mixed at the blasthole and quickly loaded into the hole (Fig. 4.14). The ratio of the amount of slurry mixed with the ANFO can be changed to offer either a higher energy load or a load which is water resistant. The cost of heavy ANFO rises with increasing amounts of slurry. The advantage, however, over cartridged products is that you fill the entire blasthole with energy and have no wasted volume which would result from cartridge loading. A disadvantage using the blends is that since the explosive occupies the entire volume of the blasthole any water in the hole is forced upward, which means that you may have to use the blend in the entire hole. Conversely with cartridge products, because of the annular space around the cartridge, you can build up to get out of water and then use the lower priced bulk ANFO.

Cartridge loading of explosives is more tedious and requires more personnel since the cartridges have to be physically taken to the blast site and stacked by each hole. The cartridges are than dropped into the borehole during the loading process.

TABLE 4.9 Characteristics of Fertilizer and Blasting Prills

	Fertilizer Prill	Blasting Prill
Density	0.9	0.8
Inert coating	3–4%	1–1.5%
Hardness	Very hard	Soft
Porosity	Nonporous	Porous
Fuel oil distribution	Surface	Throughout
Critical diameter (unconfined)	9 in.	2.5 in.

Figure 4.14   Bulk loading heavy ANFO. (Courtesy of Atlas Powder Company)

Heavy ANFO requires less personnel since explosive is pumped directly into the blasthole from the bulk truck.

Some operators try to use heavy ANFO in wet holes, however, they do not use mixtures which contain sufficient slurry. To provide the necessary water resistance, it is recommended that at least 50% slurries be used in heavy ANFO which is to be used under wet borehole conditions.

Two-Component Explosives

Two-component explosives are often called binary explosives since they are made of two separate ingredients. Neither ingredient is explosive until mixed. Binary explosives are normally not classified as explosives. They can be shipped and stored as nonexplosive materials. Commercially available, two-component explosives are a mixture of pulverized ammonium nitrate and nitromethane, which has been dyed to either a red or green color. These components are brought to the job site and only the amount needed will be mixed. Upon mixing, the material becomes cap-sensitive and is ready to use. Binary explosives can be used in applications where dynamite or cap-sensitive slurries would be used. These binary explosives can also be used as primers for blasting agents and bulk slurries. Binary explosives are not considered explosive until mixed. They, therefore, offer the small operator a greater degree of flexibility on the job. Their unit price is higher than that of dynamite. However, the money saved in transportation, magazine cost, and waste outweighs the difference in unit price. If large quantities of explosives are needed on a particular job, the higher cost per pound and the inconvenience of on-site mixing negates any savings that would be realized from less stringent storage and transportation requirements.

PROBLEMS

1. What is the critical diameter of an explosive and why is it important?
2. What is the difference between water resistance and water pressure tolerance?
3. Explosive which are not properly oxygen balanced can produce toxic fumes. Which fume(s) are produced from an explosive which has an oxygen-positive balance? Which fume(s) are produced with an oxygen-negative balance? Which fume(s) are produced if explosives get wet?
4. Define the term cycling?
5. Which class of explosives may fail to detonate reliably at temperatures below 40°F?
6. Which class of explosives is effected by temperatures above 90°F?
7. If all factors were considered equal, what difference in effect would there be on rock breakage from having one explosive detonating at 10,000 ft per sec and a second explosive detonating at 15,000 ft per sec, in equivalent borehole sizes?
8. If two explosives have identical strength ratings, does that mean that they will perform identically in the field? Explain.
9. If you use the same bulk explosive in two different diameter boreholes which are above the critical diameter, will the explosive release the same amount of energy? Explain.
10. Define loading density.
11. You will be using ammonium nitrate with a specific gravity of 0.8 in a 4 in. diameter blasthole. Determine its loading density.
12. What is the difference between a gelatin dynamite and a granular dynamite? Explain.
13. What is the difference between an emulsion explosive and a water gel explosive? Explain.
14. Define heavy ANFO.

REFERENCES

BROWN, F.W., "Determination of Basic Performance Properties of Blasting Explosives." *Quarterly (Colorado School of Mines)* 51(3):160–188.

DAMON, G.H., MASON, C.M., HANNA, N.E., and FORSHEY, D.R., "Safety Recommendations for Ammonium Nitrate-Based Blasting Agents." U.S. Bureau of Mines, IC 8746, 1977.

DICK, R.A., "The Impact of Blasting Agents and Slurries on Explosives Technology." U.S. Bureau of Mines, IC 8560, 1972, p. 44.

DICK, R.A., FLETCHER, L.R., and D'ANDREA, D.V., "Explosives and Blasting Procedures Manual," U.S. Bureau of Mines, IC 8925, 1983.

DRURY, F., and WESTMAAS, D.J., "Considerations Affecting the Selection and Use of Modern Chemical Explosives." *Proceedings of the 4th Conference on Explosives and*

Blasting Technique, New Orleans, LA, Feb. 1–3, 1979. Society of Explosives Engineers, Montville, OH, pp. 128–153.

GRANT, C.H., "Metallized Slurry Boosting: What It Is and How It Works." *Coal Age* 71, no. 4, (April 1966), pp. 90–91.

JOHANSSON, C.H. and LANGEFORS, U., "Methods of Physical Characterization of Explosives." *Proceedings 36th International Cong. of Ind. Chem.,* Brussels, 3, 1966, p. 610; available for consultation at Bureau of Mines Twin Cities Research, Minneapolis, MN.

Monsanto Co. (St. Louis, MO), *Monsanto Blasting Products AN-FO Manual* "Its Explosive Properties and Field Performance Characteristics." September 1972, p. 37.

MORHARD, R.C., (ed), "Explosive and Rock Blasting," Atlas Powder Company, Dallas, 1987, pp. 13–78.

ROBINSON, R.V., "Water Gel Explosives—Three Generations." *Canadian Min. and Met. Bull.,* 62, No. 692, (December 1969), pp. 1317–1325.

5

Initiators
and Blasthole
Delay Devices

INTRODUCTION

This chapter will give a brief overview of the available initiation systems. Since this text deals with engineering aspects of blast design, we will not go into the intricacies of hookups of the various systems. There are many texts available that go into great detail on hookup procedures.

The initiation system transfers the detonation signal from hole-to-hole at a precise time. The selection of an initiation system is critical for the success of a blast. The initiation system not only controls the sequencing of blastholes, but also affects the vibration generated from a blast, the fragmentation produced, and the backbreak and violence which will occur. Although the cost of the systems is an important consideration in the selection process, it should be a secondary consideration, especially if the most economical initiation system causes problems with backbreak, ground vibration, or fragmentation. It would be foolish to select a system based strictly on cost.

It is the intent of this section to review the currently available systems used to obtain delays in initiation both hole-to-hole within a row and row-to-row.

Initiators can be broken down into two broad classifications: electric and nonelectric. Our review of initiators will follow that sequence. The discussion will first be centered around electric methods of initiation. Electric initiation was first used by Pasley in England in 1839 for firing charges underwater.

ELECTRIC INITIATION SYSTEMS

Electric Blasting Caps

The electric blasting cap (E.B. Cap) consists of a cylindrical aluminum or copper shell containing a series of powder charges (Fig. 5.1). Electric current is supplied to the cap by means of two-leg wires that are internally connected by a small length of high-resistance wire known as the bridge wire. The bridge wire serves a function similar to the filament in an electric light bulb. When a current of sufficient intensity is passed through the bridge wire, the wire heats to incandescence and ignites a heat-sensitive flash compound. Once ignition occurs, it sets off a primer charge and base charge either near instantaneous or after traveling through a delay element which acts as an internal fuse providing a time delay before the base charge fires (Fig. 5.2). The leg wires on E.B. caps are made of either iron or copper. Each leg wire on an E.B. cap is of a different color, and all caps in a series have the same two colors of leg wires which serve as an aid in hooking up.

The leg wires enter the E.B. cap through the open end of the cap. To avoid contamination by foreign material or water, a rubber plug seals the opening so that only the leg wires pass through the plug.

There are three types of electric caps: instantaneous, millisecond delay, and long period delay caps. Instantaneous caps fire within a few millisecond (ms) after they receive the firing current. The millisecond delay series caps sequence in fixed increments of milliseconds such as 25 ms, 50 ms, or 100 ms. Long period delays fire in time increments of 0.25 to 1.5 sec.

Because of the recognized importance of having an accurate safe initiation system, many companies are in the process of research and development of new electric initiation systems. One system is the Magnadet system of initiation which was invented by ICI in Scotland.

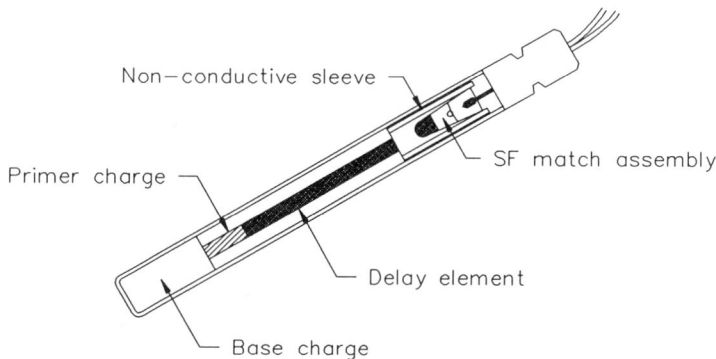

Figure 5.1   Delay electric blasting cap. (Courtesy of Atlas Powder Company)

Figure 5.2   Electric cap during detonation.

Magnadet electric detonator and magna primer.   The transfer of electrical energy in the Magnadet system is not by direct wiring connections. The system functions by electromagnetic induction between the primary and secondary coil of a transformer (Fig. 5.3).

An AC power source operating at a frequency of 15,000 Hz or above is provided by a special blasting machine.

Magnadet consists of a separate transformer external to each detonator. The transformer device is 20 mm outside diameter, 10 mm inside diameter by 10 mm wide ferrite ring. Lead wires from each detonator are attached to its own ferrite ring and form the secondary windings of the transformer. A plastic covered connecting wire passing through the center of each ferrite ring forms the primary winding of the transformer. The detonator portion of the assembly is of conventional construction. The ferrite ring with the secondary windings leading to the detonator wire is encapsulated within a plastic sheath for protection against damage. Each plastic protector is stamped with a number corresponding to the delay number of the detonator to which it is attached. Special sequential blasting machines are available to work with Magnadet caps.

The magna sliding primer is a cast pentolite booster within a specially designed plastic housing that can accommodate Magnadet electric caps with 50 mm leads. The central hole allows a length of electrical cable to be threaded through the primer and through the ferrite rings of the Magnadet electric detonators, thus providing the primary circuit. An inductive coupling is formed as previously explained (Fig. 5.4).

If more than one magna primer is required per hole, such as when firing

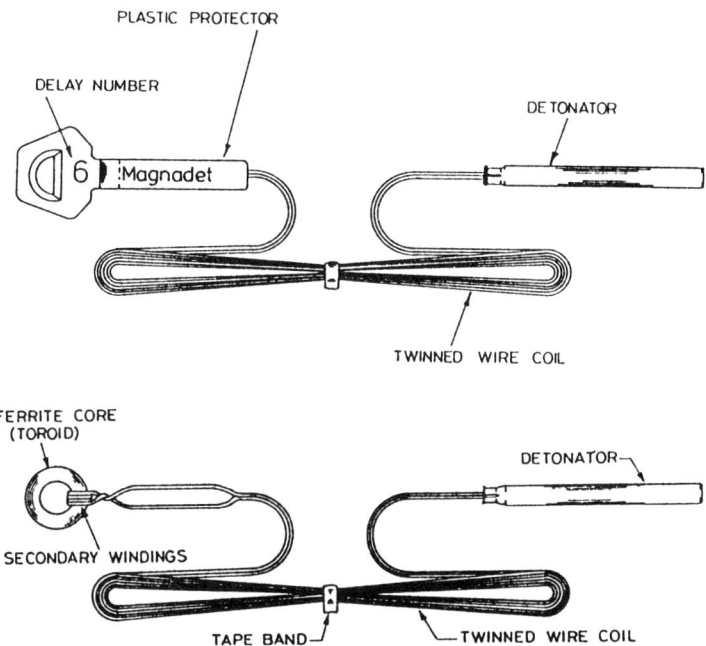

PLASTIC PROTECTOR

DELAY NUMBER

DETONATOR

Magnadet

TWINNED WIRE COIL

FERRITE CORE
(TOROID)

DETONATOR

SECONDARY WINDINGS

TAPE BAND

TWINNED WIRE COIL

Figure 5.3 Magnadet electric cap.

decked charges, then subsequent magna primer delay units can be slid down on the same primary cable to the desired location within the hole (Fig. 5.5).

The Magnadet system provides safety advantages where electrical hazards are present. It offers protection against stray currents from DC power sources since the transformer device will not respond to DC energy. Stray currents from AC power sources is too low a frequency, 50 or 60 Hz compared to 15,000 Hz required for reliable firing. The assembly is designed to withstand static electricity hazards associated with pneumatically loaded ANFO. It offers protection against radio frequency energy. The effective voltage across each unit is low. Typically one to two volts, therefore, is insufficient voltage for current leakage to occur.

The system can result in appreciable time saving in the loading and connecting procedure. All that is required is to thread a wire continuously through the hole in the plastic protector attached to the cap leg wires protruding from each hole.

Magna primers offer a sliding primer system which allows firing decked charges with each deck fired on a separate delay.

Electronic blasting caps.   A definite need has surfaced for super-accurate delays. Electronic technology has advanced to the point that technology exists to create electronic delays at a reasonable cost. In some countries, electronic delays are already being used. An electronic detonator with super-accurate firing

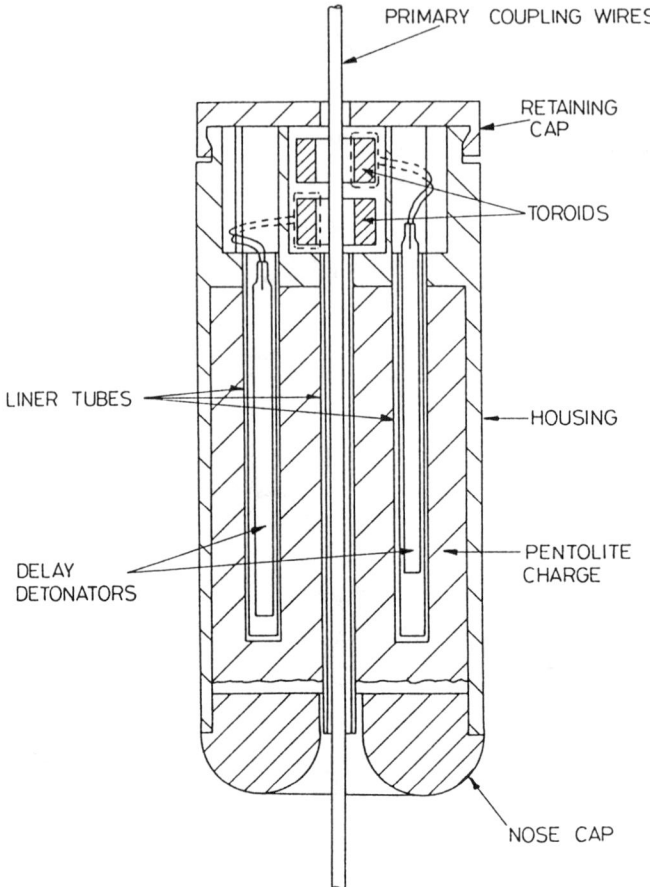

PRIMARY COUPLING WIRES

RETAINING CAP

TOROIDS

LINER TUBES

HOUSING

DELAY DETONATORS

PENTOLITE CHARGE

NOSE CAP

Figure 5.4   Magna primer.

times and the ability to have infinite delay periods at any interval of time will revolutionize the blasting industry. This initiation system would virtually elimi-nate large tolerances in firing time and better fragmentation will result. Caps could be given a specific code, whereby accidental firing would not be a hazard.

Sequential Blasting Machine

The sequential blasting machine was developed by Research Energy of Ohio, Inc., as a solid state condenser discharge blasting machine with a sequential timer that permits the detonation of many electric caps, 175 ohms per circuit, at different, precisely timed intervals. The machine consists of ten different firing circuits that are programmed to fire one after another at selected intervals. The combination of

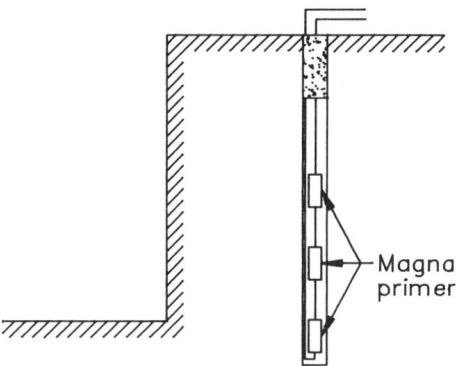

Figure 5.5   Magna sliding primer.

ten different circuits, or intervals, in conjunction with delay blasting caps can yield many independent delays within the blast.

Timing devices which provide additional delays for electric caps have been available for decades but were not widely used. Early mechanical machines with moving parts caused misfires when contact points became dirty or worn. Solid state units eliminated the problems with misfires, however, they did not eliminate the inconvenience of bringing the leads of each cap back to the timer unit.

The sequential blasting machines reduced the hookup time by timing series of caps rather than individual caps.

Sequential timers are used in construction as well as mining applications. The timers allow the use of many delays within a blast. Therefore, the pounds of explosives fired per delay period can be reduced to control noise and vibration effects. The sequential blasting machine can be set to fire from 5 to 199 ms in increments of 1 ms (Fig. 5.6).

NONELECTRIC INITIATION SYSTEMS

Nonelectric initiation systems have been used in the explosive industry since blasting began. The fuse was the first and oldest method of nonelectric initiation. It provided a low cost, but a hazardous system. An improved safety fuse was introduced in 1831 by Bickford. The fuse would not reliably initiate high explosives, and blasting caps were used in conjunction with the safety fuse. The cap and fuse system has declined in use with the introduction of more sophisticated, less dangerous methods. Accurate timing with cap and fuse is impossible. The system has no place in a modern blasting industry because of the inaccurate timing and hazards.

Four nonelectric initiation systems are available. All may find use in the blasting industry. Blasters often combine the use of more that one nonelectric

Figure 5.6   Sequential timer. (Courtesy of Research Energy of Ohio, Inc.)

system on a blast to increase the number of delays available. Often electric and nonelectric system components are combined to give a larger selection of delays and specific delay times.

Detaline Initiation System

The Detaline system manufactured by ETI is a two-path nonelectric system compatible with detonating cord downlines and nonelectric in-hole delays. The Detaline system consists of Detaline cord, Detaline starter, Detaline ms surface delays, and Detaline ms in-hole delays (Fig. 5.7).

Detaline cord is a low energy detonating cord having a pentaerythritoltetranitrate (PETN) explosive charge of 2.4 grains of explosive per foot. The explosive core is within textile fibers and covered by a seamless outer plastic jacket. Detaline cord will not propagate through a knotted splice. To splice the cord, a Detaline starter is needed. A starter is also needed to initiate the Detaline trunkline.

The Detaline starter is in the shape of an arrow to designate the initiation direction when hooked up. The starter consists of an aluminum cylinder containing

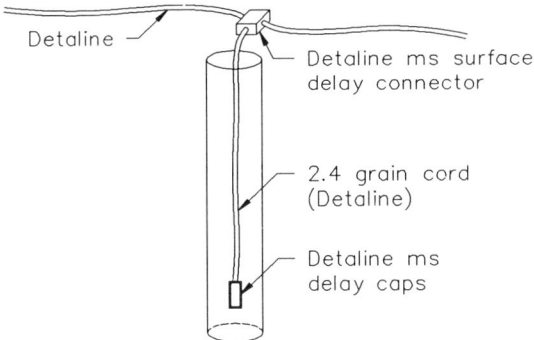

Detaline

Detaline ms surface
delay connector

2.4 grain cord
(Detaline)

Detaline ms
delay caps

Figure 5.7 Detaline system with components.

an instantaneous explosive charge. The Detaline cord is bent into a V-shape, inserted into the starter, and locked in place with the sawtooth pin provided.

Detaline ms surface delays are shaped like the Detaline starter and contain an element that provides a time delay between activation and initiation of the detonating cord downline locked in the pointed arrow end.

Detaline ms in-hole delays resemble an ordinary blasting cap except for a special top closure that is designed for insertion of a Detaline cord. A delay tag is affixed to the shell of each cap for delay period identification.

Detaline systems are connected similar to conventional detonating cord systems except that connections are made easier and no right angle connections are necessary. The system is connected to create a redundant, two-path system. A cap and fuse or an electric blasting cap is inserted into a starter to initiate the Detaline trunkline.

Detonating Cord and Compatible Delay Systems

Detonating cord is a round, flexible cord containing a center core of high explosive, usually PETN, within a reinforced waterproofing covering. Detonating cord is relatively insensitive and requires a proper detonator, such as No. 8 strength blasting cap for initiation. It has a very high velocity of detonation approximately 21,000 ft/sec. The cords detonation fires cap-sensitive high explosives with which it comes into contact, also adjoining detonating cord, and nonelectric delay caps. Detonating cord is insensitive to ordinary shock and friction. Surface as well as in-hole delays can be achieved by proper delay devices attached to detonating cord. A major disadvantage in the use of detonating cord on the surface is the loud noise produced as the cord detonates. Grass and brush fires have been started by using detonating cord in dry areas.

The Ensign Bickford Company produces delay connectors for use with detonating cord. They consist of two molded plastic units which contain an aluminum tube delay element in the center portion. The two relay elements are connected with an 18–in. length of Nonel tubing. Each end of the unit is made so that the detonating cord can be looped and locked in the connector.

The assembly is bi-directional and each delay element fires only in one direction. They are installed by cutting the detonating cord and attaching the ms connector units to the cut ends.

Down-hole delay devices.   In-hole delay can be achieved with detonating cord using Primaline Primadets which consists of a 4 grain per foot miniaturized detonating cord connected to a delay blasting cap. Primaline primadets should not be used with cap-sensitive explosives because the cord can initiate the explosive and bypass the delay. The SM Series combines the older long period and millisecond series into one sequence of 17 delays. SM Series primadet consists of a primaline, a high strength blasting cap, a delay tag and a connector to secure it to detonating cord surface lines. Primaline Primadets loaded in blastholes are initiated by a Primacord trunkline.

Delayed primers are units that contain a cap-sensitive high-explosive primer with a nonelectric delay cap held in a detonation relationship with a downline of detonating cord (Fig. 5.8). The delayed primer is used for detonating ANFO, blasting agents, or any non-cap-sensitive explosives. They are ideal for bottom or top initiation and can be used for either full column or deck loaded holes.

A tube runs alongside the primer in which a single downline is threaded. This eliminates the need of separate downline within each individual deck. Any number of primers can be threaded on this single downline. The delay inserts (Fig. 5.9) (shipped separately and field assembled) are in 16 periods with millisecond designation tags.

A 25 grain detonating cord downline that has a tensile strength in excess of 200 lb is recommended for use with delayed primers.

Hercudet System

Hercudet is a noiseless, nonelectric initiation system in which the ignition charge in each individual cap is activated by a gas detonation traveling through plastic tubing from cap-to-cap at a speed of 8,000 ft/sec.

Figure 5.8   Austin delayed primer. (Courtesy of Austin Powder Company)

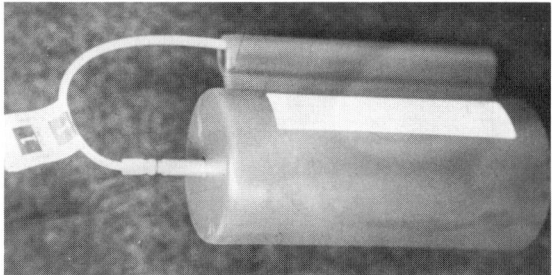

Figure 5.9   Components of delay primer. (Courtesy of Austin Powder Company)

The Hercudet cap is a 4 in. cap of Number 10 strength (Fig. 5.10). This cap has an empty air space above the ignition charge instead of the usual bridge wire associated with an electric cap. Other than that, the cap components are similar to electric blasting caps.

In order to ready this cap for use, it must be connected in such a way that there is an open piping-like path from one end of the series of caps to the other (Fig. 5.11). A number of such series may then be connected in parallel with one another so that a gas, introduced to a common trunkline inlet tube, will travel from that tube into each series of caps and out the open end of each of the series. This path is first used for testing the circuit with air or nitrogen. After testing, when the area is ready for blasting, a explosive mixture of gaseous oxygen and fuel is introduced and initiated by the Hercudet blasting machine. The gas detonation starts the ignition charge in each cap. The reaction through the tubing is virtually noiseless. The detonation rate of the mixed gas is 8,000 ft/sec. At 8,000 ft/sec, it takes 10 ms to traverse through 80 ft of tubing. This tubing delay can often be used to supplement the built-in delay element in the cap. The propagation speed of gas detonation in the tube provides additional delay, thus, different delay action is achieved through the combining of tubing delay and cap delay.

You can use cap-sensitive explosive in the blasthole since it cannot be initiated by the tube in the hole. Dead-pressing or densitizing of explosives cannot occur.

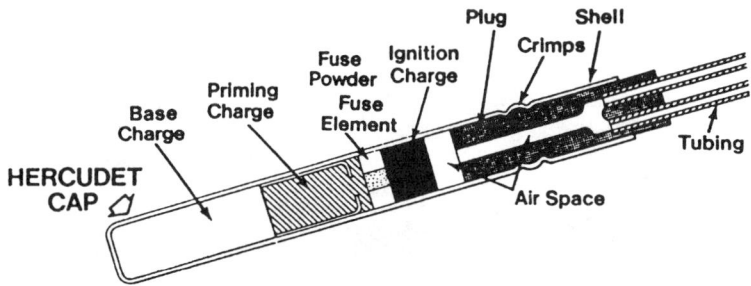

Figure 5.10   Components of Hercudet cap. (Courtesy of IRECO)

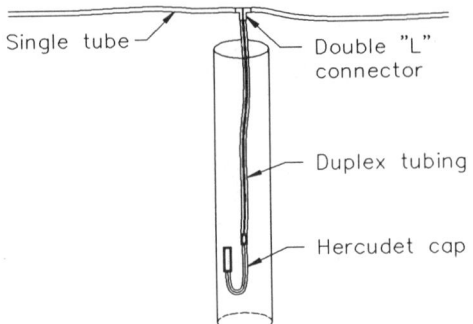

Single tube

Double "L"
connector

Duplex tubing

Hercudet cap

Figure 5.11 Hercudet circuits with
components.

The circuit test capability allows checking for proper hookup.

The inert until charged characteristic provides a safety advantage by eliminating all electrical hazards.

Shock Tube Systems

The Nonel shock tube is a nonelectric, instantaneous, nondisruptive signal transmission system. A similar system is also manufactured by Atlas Powder Co. and is called Blastmaster. Blastmaster functions similar to Nonel. Nonel consists of a .125 in. diameter plastic tube made of Surlyn and has a thin coating of reactive material on the inside (Fig. 5.12). This reactive material has a powder weight of .1 grains per ft (1/70000 lb) and propagates a noiseless shock wave signal at a speed of

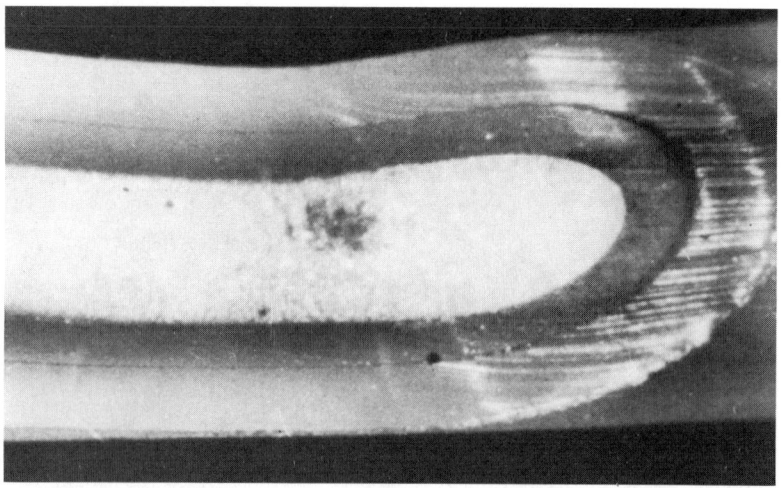

Figure 5.12 Nonel shock tube. (Courtesy of Ensign-Bickford Company)

6,000 ft per sec. The system eliminates all electrical hazards except possible initiation by direct lightning strike.

Nonel may be initiated by detonating cord, electric blasting cap, cap and fuse or a Nonel starter which initiates with a shotgun primer. It is safe from most electrical hazards and it is noiseless on the surface.

Nonel primadets are suited for use with commercially available dynamites or cap-sensitive water gel or emulsion-type high explosives because the tube will not initiate or desensitize these explosives. Nonel primadets can be used for initiation with a suitable primer.

Millisecond (ms) and Long Period (LP) series Nonel primadets provide non-electric delay blasthole initiation. The assembly consists of a Nonel shock tube, a high-strength blasting cap, and a delay tag for identification (Fig. 5.13). The blasting cap will initiate ANFO in dry hole up to 3 in. in diameter without a primer.

A rugged model of the millisecond series called the H.D. Primadet is designed for open pit, strip mine, quarries, and construction. H.D. Nonel Primadets consist of a length of Nonel (15 in.) tube which is heat sealed on one end and has a millisecond delay blasting cap crimped to the open end. This unit is field assembled to the proper lead length desired by tying to a specially designed 7.5 grain detonating cord (Fig. 5.14). The small explosive charge in the detonating cord allows it to function with minimum disruption of ANFO, slurry or other non-cap-sensitive explosives. At the surface, the Primaline is attached to the Primacord trunkline. H.D. Primaline is tied to the H.D. Primadet with a square knot.

Long length, heavy duty (L.L.H.D.) primadets are similar to the H.D. Pri-

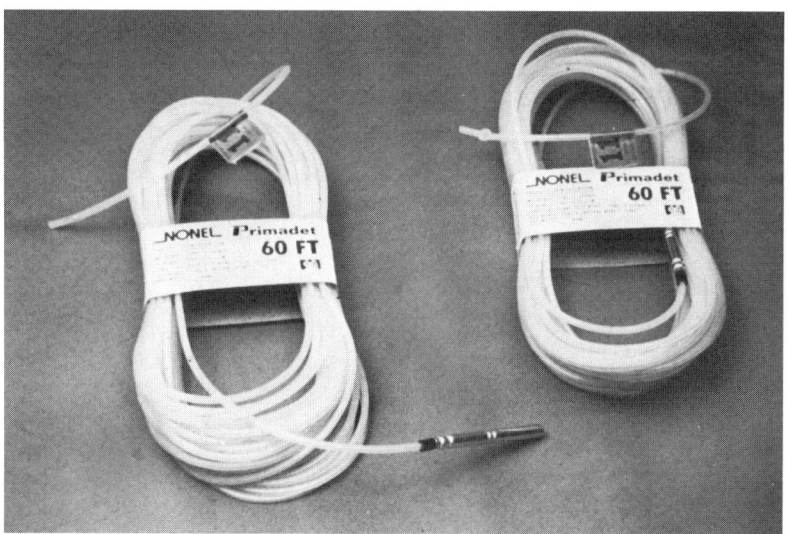

Figure 5.13   Nonel primadet. (Courtesy of Ensign-Bickford Company)

With 7 grain detonating cord

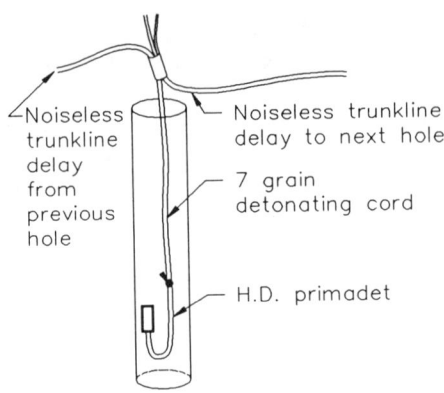

Without detonating cord

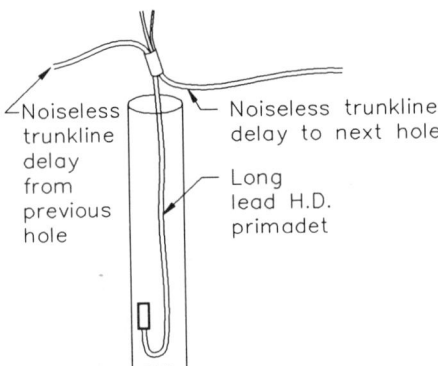

Figure 5.14 Nonel hook-up with and without detonating cord.

madet except rather than using a 15–in. H.D. Primadet which is connected to 7.5 gram Primaline, the L.L.H.D. unit has a long length Nonel tube which extends to the collar of the blasthole. The long length Nonel tube eliminates the need for detonating cord in the blasthole which allows the use of cap-sensitive explosives in the hole. L.L.H.D. primadets are available in the same periods as regular H.D. Primadets.

Noiseless trunkline delays are used in place of Primacord trunklines. All units contain built-in delays to replace conventional MS connectors used with detonating cord. Trunkline delays are factory assembled units with five main components: the Nonel shock tube, the blasting cap, the connector, the delay tag, and the plastic sleeve (Fig. 5.15).

Noiseless Nonel lead-in line is used as the nonelectric primary initiator for the blast. Nonel lead-in consists of a continuous length of Nonel tube heat sealed on one

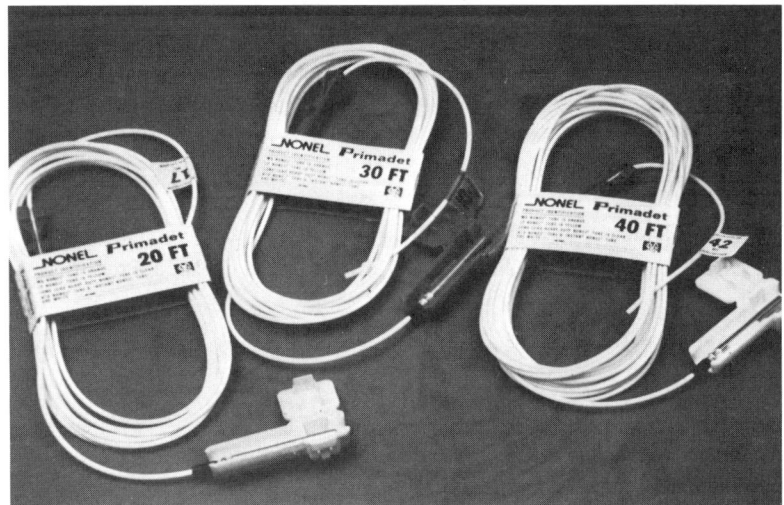

Figure 5.15   Nonel surface delay. (Courtesy of Ensign-Bickford Company)

end with an instantaneous blasting cap on the opposite end. The Nonel lead-in cap end is attached to the downline of the first hole to fire in the blast. Noiseless lead-in lines may be initiated by electric blasting caps, cap and fuse, or a Nonel starter. Nonel lead-in lines are available on spools in continuous lengths of 200, 500, and 1000 ft.

PROBLEMS

1. How is the delay achieved in delay electric blasting caps?

2. How does the Magnadet electric cap differ from a standard electric blasting cap?

3. Describe the advantages of using a sequential blasting machine.

4. How is the transfer of detonation between a Detaline surface line and a Detaline downline achieved?

5. What is a delay primer?

6. How is the transfer of detonation accomplished from hole-to-hole with the Hercudet? How is the transfer of detonation from hole-to-hole achieved with the Nonel system?

7. What is the reaction rate of Hercudet?

8. What is the reaction rate of Nonel?

9. What is the reaction rate of Detonating Cord?

REFERENCES

Blasters Handbook, E.I. du Pont, de Nemours & Co., Wilmington, 1977.

DICK, R.A., FLETCHER, L.R., and D'ANDREA, D.V., "Explosives and Blasting Procedures Manual," U.S. Bureau of Mines, IC 8925, 1983.

Handbook of Electric Blasting, Atlas Powder Co., Dallas, 1985.

"Internationally Accepted Standards and Sequential Blasting Control Systems and Accessories," Research Energy of Ohio, Huron, 1988.

MORHARD, R.C., (ed), "Explosive and Rock Blasting," Atlas Powder Company, Dallas, 1987, pp. 79–156.

Shotfires Guide, Hercules, Inc., Wilmington, 1978.

"Technical Bulletins for Nonel Primadets, Primacord and Related Products," Ensign-Bickford Co., Simsbury, CT, 1985.

6

Primer and Booster Selection

INTRODUCTION

Differences Between Primers and Boosters

Cap sensitive.   Explosives will reliably initiate with only the energy from a blasting cap. Blasting agents and some slurries cannot reliably initiate with the energy released from the cap. These products obtain the additional energy for initiation from a primer.

The difference between a primer and booster is in its use, rather than in its physical composition or makeup. A *primer* is defined as an explosive unit which contains an initiator. As an example, if a blasting cap is placed into a cartridge of dynamite, that cartridge with initiator becomes the primer. A *booster,* on the other hand, is an explosive unit of higher energy than the borehole charge and does not contain a firing device. The booster is initiated by the column charge adjacent to it. A booster is used to put additional energy into a hard or tough rock layer (Fig. 6.1).

Before we can select the primer we must understand the following:

1. The effect of primer explosive composition
2. The effects of primer diameter
3. The influence of primer length
4. The use or need of multiple primers.

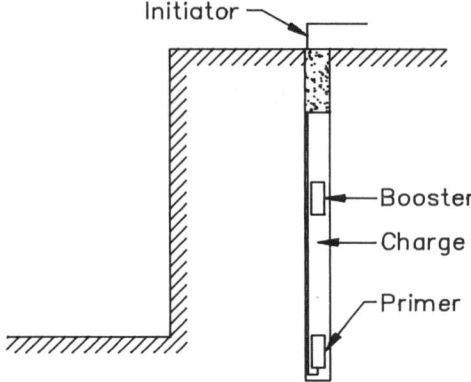

Figure 6.1 Difference between a primer and a booster.

Primer Types

Primers can be found in many sizes and in many varying compositions. Primers may be as small as a Detaprime, which looks similar to a wax sleeve which fits over the end of a blasting cap and weighs a few ounces, or may consist of a 50 lb cartridge of explosive. Primer diameters can vary from a fraction of an inch to well over a foot. Primers come in many different compositions. Various grades of dynamite are used as primers as well as water gels, emulsions, and densified ammonium nitrate compounds. Various types of cast explosives of high density and velocity are also used for priming (Fig. 6.2). The vast number of sizes and compositions of primers can be confusing for the operator and improper selections are often made which can produce poor results.

Inadequate priming can be costly. If the main charge in a borehole is not being

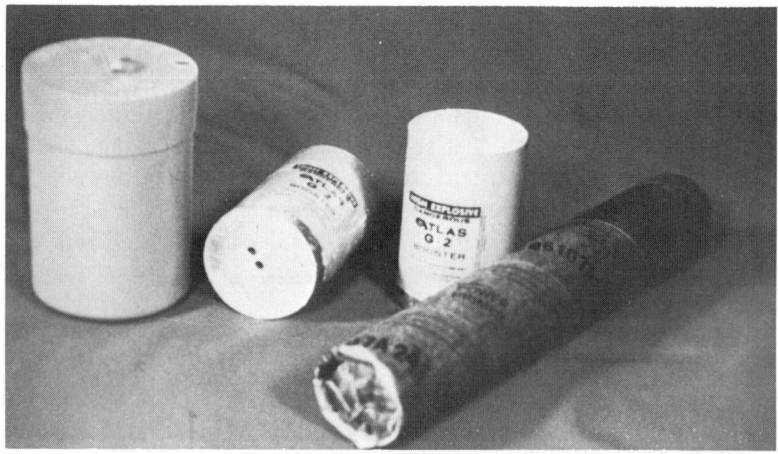

Figure 6.2 Primers of different sizes. (Courtesy of Atlas Powder Company)

properly primed, patterns by necessity may be much smaller than would be used with good priming procedures. Fragmentation size may also get larger. Poor priming procedures are not only costly, but can cause excessive ground vibration, airblast, flyrock and more damage behind the last row of holes than would occur if good procedures were used.

Multiple primers. The number of primers that are placed in a blasthole is dependent on a number of different factors. There is no one method of priming which would define a universally accepted procedure.

Routinely putting two primers into a blasthole regardless of the borehole length is common practice for some operators. They are concerned about the possibility of getting a poor blasting cap, which may not fire, or are concerned about *cutoffs* of the hole because of shifting rock caused by the firing of a hole on an earlier delay. In either case, their rationale is that using a second primer is insurance against problems. If a rock mass contains mud seams, the main charge may have insufficient confinement which could be a cause for concern. With these geologic conditions present, it is common to find operators placing additional primers in the blasthole to cause the explosive charge to fire more rapidly, thereby reducing possible problems because of loss of confinement at the seams. If the blaster is working in competent rock, the use of blastholes whose length is greater than twice the burden may require a second primer to get efficient detonation throughout the total length of the charge. Conversely, in most cases from a purely technical standpoint, only one primer is needed for a single column charge of explosive especially if the bench height is less than twice the burden. In these cases where more than one primer is used, it would be necessary that both primers would be firing near instantaneously.

If two or more primers are being placed in a blasthole as insurance against cutoffs, the second primer should be placed on a later delay period. The location of the first primer may be critical for the shot to perform properly. The second delayed primer, therefore, would act only as a back-up unit should the first one fail to initiate at the proper time. A method to calculate the number of primers needed in a blasthole is given in Chap. 7.

Primer characteristics. Primer composition and primer size are the two most critical criteria in primer selection. The primer composition determines the detonation pressure which is directly responsible for the initiation of the main charge. Research conducted by Junk (1968) demonstrated that primer composition significantly affected the performance of ANFO charges. Figure 6.3 illustrates the effect of using different primers of the same size in a 3–in. diameter ANFO charge and the response of the ANFO at various distances from the primer. Primers with low-detonation pressure cause a burning reaction to start rather than a detonation. All primers producing detonation velocities above steady state would be acceptable. Steady state is defined as the normal velocity at which the explosive would detonate for the given diameter (Fig. 6.4).

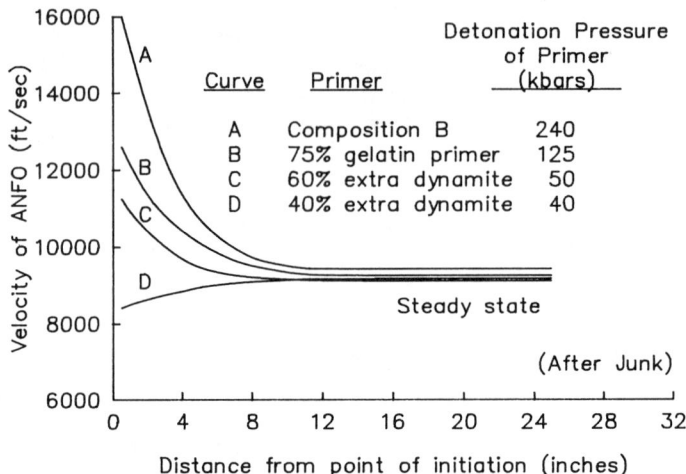

Figure 6.3   Explosive composition and primer performance.

Primer size is also important to obtain a proper reaction. Small diameter primers are not as efficient as large diameter units. Figure 6.5 demonstrates the effect of primer diameter on ANFO response in 3-in. diameter charges at various

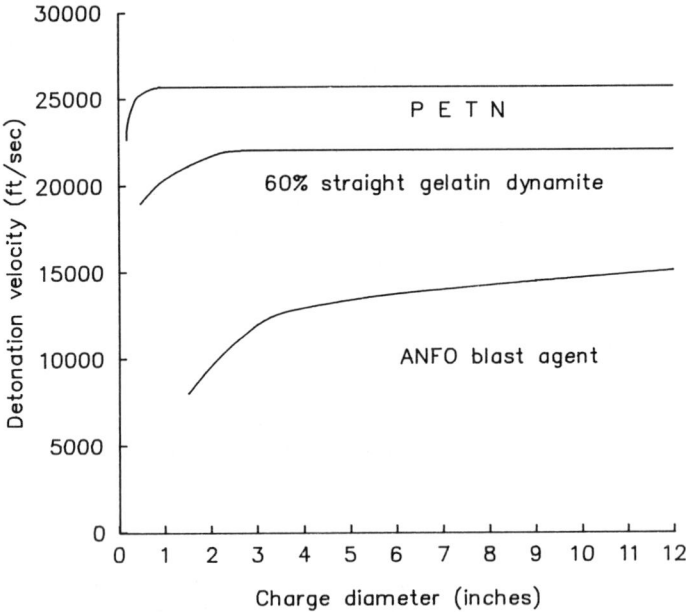

Figure 6.4   Steady state velocity for different products.

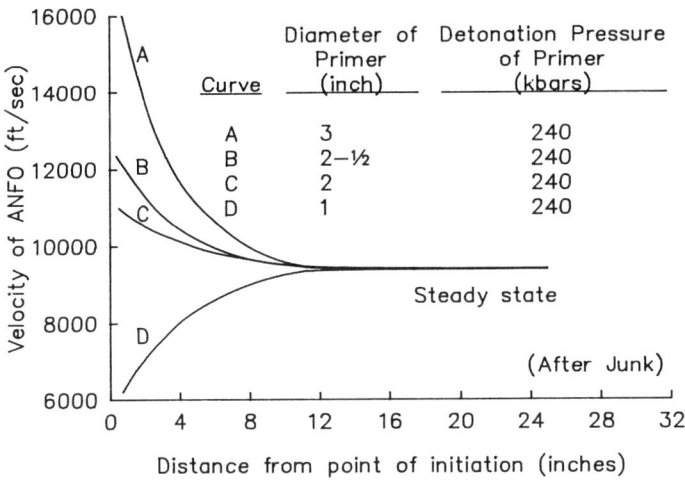

Figure 6.5   Primer diameter and charge performance.

distances from the primers. For these tests all conditions were held constant except primer diameter. The one-inch diameter primer started the ANFO at a velocity of 6000 fps which was below steady state. The results of this research showed that small diameter primers become inefficient regardless of the composition of the material used.

The necessary primer length or number of priming cartridges used is commonly selected at random without understanding the function of the length. The energy component which is used to transfer the detonation from the primer to the main charge is the shock energy. Shock energy is kinetic energy and is proportional to the detonation velocity. Once the primer cartridges reach maximum velocity it cannot produce more shock energy regardless of additional length, therefore, the primer cartridges need to be of sufficient length to reach its maximum velocity.

Primer selection guidelines.   The following are some general guidelines for priming:

1. The detonation pressure of a primer must be above the level necessary to cause the column charge to detonate at or above steady state velocity. To accomplish this, the specific gravity and confined detonation velocity can be used as indicators of detonation pressure. A primer that would have a specific gravity of approximately 1.2 with a confined detonation velocity greater than 15,000 ft per sec would normally be adequate when priming non-cap-sensitive explosives, materials such as ANFO, blasting agents and water gels.

This combination of density and velocity produces a detonation pressure of about 60 kilobars. On the other hand, for explosives such as emulsions which would detonate at higher reaction rates, more energetic primers would produce better results. In these cases, a primer with a specific gravity of 1.3 with a confined detonation velocity greater than 17,000 ft per sec would be adequate to more quickly achieve the explosive columns steady state reaction velocity and maximum energy release. This combination of density and velocity produces a detonation pressure of about 80 kilobars.

2. The diameter of the primer should be larger than the critical diameter of the explosive used for the main column charge.

3. The primer must be sensitive to the initiator. A wide variety of products are used as primers. These primers have different sensitivities. Some may be initiated by low energy detonating cord, while others may be insensitive to these initiators. It is important that the blaster understand the sensitivity of the primer to ensure that detonation in the main column charge will properly occur.

4. The explosive in the primer must reach its rated velocity of detonation within the length of the cartridge. If this is achieved then, additional cartridges of primer explosive are unnecessary and serve no useful purpose.

5. For most blasting applications no more than two primers per blasthole are needed. The second primer although technically not needed is commonly used as a backup system should the first primer fail or fail to shoot the entire charge.

Booster

Boosters are used to intensify the explosive reaction at a particular location within the explosive column. Boosters are sometimes used between each cartridge of detonating explosive to ensure a detonation transfer across the ties of the cartridge. This normally is a poor excuse for the use of boosters, since the price of boosters is greater than the explosive cost per pound in the main charge. The selection of an explosive charge in a cartridge which would not require a booster between each cartridge may be a more economical solution.

In general, boosters are used to put more energy into a hard layer within the rock column. They are also used to intensify the reaction around the primer which will put more energy at the primer location. They are commonly used when primers are near the bottom of the hole, because the bottom of the hole is the hardest place to break. Using a booster at hole bottom normally allows an increase in the burden dimension and better breakage at floor level. Boosters can be made of similar explosive materials as primers. Their sole function is to place more energy at point locations within the explosive column.

Effects of Detonating Cord on Energy Release

Cap sensitive explosives such as dynamite are initiated by detonating cord. Non-cap-sensitive explosives such as ammonium nitrate, emulsions, and water gels can be effected in many different ways by detonating cord passing through the powder column. If the detonating cord has sufficient energy and size some explosives may detonate or burn. A burning reaction rather than a detonation releases only a small fraction of the explosives available energy. The blast is underloaded because of a low energy release and ground vibration levels increase while blast holes may vent and produce flyrock.

To prevent the main explosive charge from burning or deflagrating we must be sure that the detonating cord is not too large for the borehole. Cord grain loads that should not cause deflagration are as follows:

Borehole Diameter (inch)	Maximum Cord Load (grains/foot)
2–5	10
5–8	25
8–15	50

If the detonating cord is not of sufficient size to cause a reaction in the explosive, it can cause the explosive to be damaged. The location of the cord can be in the center, or side of the hole and its location will control the severity of affects. The damage that results is called dead pressing or precompression. *Dead pressing* increases the explosive density and it will not detonate. This occurs when the detonating cord is of sufficient energy to crush out the air spaces within the explosive or to break the air filled microspheres placed in some products which also provides air to form hot spots for detonation. The adiabatic compression of air is necessary for detonation to proceed throughout the explosive.

When the explosive is only partially compressed or damaged by precompression it may detonate or burn releasing only a fraction of the available energy. This effect can be confusing since the explosive is totally consumed yet little rock breakage results. Commonly, the blaster who suffers this type of problem thinks that the problem is because of hard, tough rock. We can obtain a better understanding of this problem as we look at the energy loss which results from passing a detonating cord through an explosive column. Figure 6.6 shows the energy loss for ANFO, which is damaged by detonating cord. Slurry can also suffer similar damage. The figure shows that even a four-grain detonating cord can cause a significant energy loss in ANFO. Approximately 38% of the useful energy is lost with as little as a four-grain cord in a two-inch diameter blasthole.

The general recommendation is not to use any detonating cord in small diameter holes.

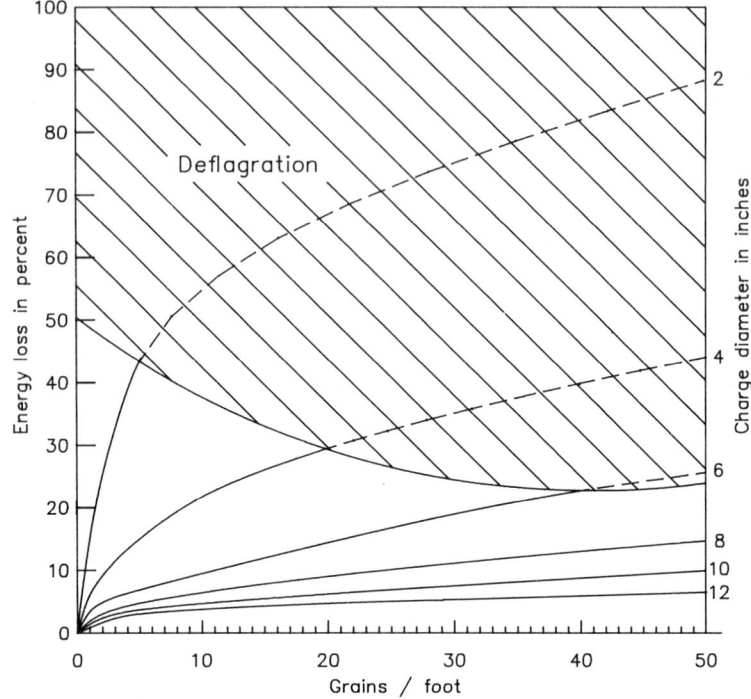

Figure 6.6   Energy loss caused by detonating cord.

PROBLEMS

1. What is the difference between cap sensitive and non-cap-sensitive explosives?
2. What is the difference between a primer and a booster?
3. What are the two most important criteria used in selecting an explosive as a primer cartridge?
4. What length of primer cartridge is needed to be sure that it will function properly?
5. What effects can detonating cord have on small diameter non-cap-sensitive explosives? Explain.
6. Define the term precompression and explain its importance.
7. What is meant by the term steady state velocity?

REFERENCES

BHUSHAN, V., KONYA, C.J., and LUKOVIC, S., "Effects of Detonating Cord Downline on Explosive Energy Release." *Proceedings of the Second Mini-Symposium on Explosives and Blasting Research,* Society of Explosive Engineers, Montville, OH, 1986, pp. 41–55.

CONDON, J.L., and SNODGRASS, J.J., "Effects of Primer Type and Borehole Diameter on An-Fo Detonation Velocities." *Min. Cong. J.*, 60, no. 6, (June 1974), pp. 46–47, 50–52.

DICK, R.A., "Puzzled About Primers for Large Diameter An-Fo Charges? Here's Some Help to End the Mystery," *Coal Age*, 81, no. 8, (August 1976), pp. 102–107.

HAGAN, T.N., "Optimum Priming for Ammonium Nitrate Fuel-Oil Type Explosives," *Proc. Southern and Central Queensland Conf. of the Australasian Inst. of Min. and Met.*, Parkville, Australia, July 1974, pp. 283–297, available for consultation at Bureau of Mines Twin Cities Research Center, Minneapolis, MN.

KONYA, C.J., "Directional Effects of Small Diameter Primers," *Proceedings of Sixth Conference on Explosive and Blasting Technique*, Tampa, FL, 1980.

KONYA, C.J., "Priming and Boostering Practices," *Proceedings of Explosives and Blasting Conference*, Lexington, KY, 1974.

KONYA, C.J., and FOLDESI, J., "As Inicialasi Pontok Szamanak Meghatarozasa Ando—val Toltott Robbantolyukak Eseten," Epitoanyag, Budapest, Hungary, December 1975.

KONYA, C.J., and FOLDESI, J., "Priming Techniques Employed at the Tallya Quarry," *Proceedings of the Second Conference of Explosives and Blasting Technique*, Louisville, KY, February 1976.

JUNK, N.M., "Research on Primers for Blasting Agents," *Min. Cong. J.*, 50, no. 4, (April 1964), pp. 98–101.

JUNK, N.M., "The Principles of Priming and Boostering An-Fo with Slurry Explosives," Preprint No. 68-F-7, Annual Meeting AIME, 1968.

7

Blasthole Design

INTRODUCTION

The design of any blast must encompass the fundamental concepts of ideal blast design, which are then modified when necessary to account for local geologic conditions. A plan must be designed one step at a time. This chapter will discuss a step-by-step procedure for the development of a blasting plan. Methods to determine whether design variables are in normally acceptable ranges will also be discussed.

During the design of any blasting plan, the engineer must select the proper variables to match the specific field conditions. There are two types of variables that are commonly discussed in blasting. Those variables over which we have little control such as the geology, rock characteristics, and regulations or specifications as well as the distance to the nearest structures are uncontrollable variables. These uncontrollable variables require the blaster to take a standard design and modify it to fit within the constraints of the job. To overcome the limitations posed by uncontrollable variables, the blaster has design factors at his disposal which he can use to his benefit. The variables over which we do have control are called controlled variables. Some examples are

1. hole diameter
2. hole depth
3. subdrilling depth
4. stemming distance
5. stemming material

6. spacing
7. number of holes in the blast
8. direction of rock motion
9. deck loading, (if necessary).

Along with these physical dimensions, we also have control over time. Time must be scaled and is considered just as important as the physical dimension. To control time, the blaster can

1. select blasthole timing to fit the blast geometry
2. select an initiation system which will produce the proper time sequences
3. plan for cap scatter in the blasting pattern
4. design a pattern to control the vibration at protected structures
5. select the proper time to ensure maximum fragmentation
6. select the proper time to ensure the minimum airblast
7. select the proper time for wall control
8. select the proper time to reduce flyrock.

In order to produce the proper results, the blaster must also consider the explosives supplies available to him such as

1. the use of an initiation system which is safe and reliable
2. the type of explosive which will properly perform in the specific environment
3. the explosives delivered energy
4. whether to bulk or cartridge load the blasthole
5. how to best handle water problems.

Burden

Burden distance is defined as the shortest distance to relief at the time the hole detonates (Fig. 7.1). Relief is normally considered to be either a ledge face or the internal face created by a row of holes that have been previously shot on an earlier delay. The selection of the proper burden is one of the most important decisions made in any blast design. Of all the design dimensions in blasting, it is the most critical. If burdens are too small, rock is thrown a considerable distance from the face. Airblast levels are high and the fragmentation may be excessively fine. If burdens are too large, severe backbreak occurs behind the last row of holes and back shattering results on the back wall. Excessive burdens may also cause blastholes to geyser causing flyrock. Vertical cratering and high levels of airblast will also occur. Excessive burdens cause overconfinement of the blastholes, which result in high levels of ground vibration per pound of explosive used. Rock breakage can be extremely coarse and bottom or toe problems often result. Of all the design vari-

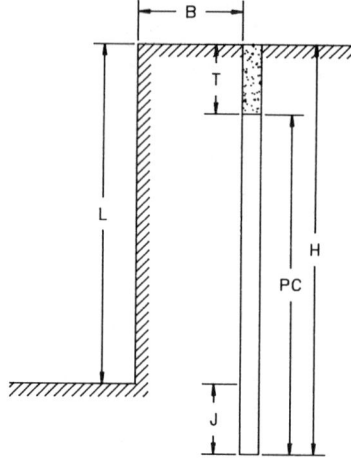

L = Bench height
B = Burden distance
T = Stemming distance
J = Subdrill distance
PC = Powder column length
H = Blasthole length

Figure 7.1   Symbols used for blast design.

ables, there is the least allowable error in the burden dimension. Other variables are more flexible and will not produce the drastic differences in results as would the same proportion of error in the burden dimension.

If the operator has selected a burden and used it successfully for a drill hole of another size and wants to determine a burden for a drill hole that is either larger or smaller, the operator can do so quite easily if the only thing changed is the size of the hole, and the rock type and explosives are unchanged. To do this, use the following simple ratio:

$$B_2 = B_1 De_2/De_1 \tag{7.1}$$

where:

B_1 = burden successfully used on previous blasts
De_1 = diameter of charge for B_1
B_2 = new burden
De_2 = new explosive diameter

Example 7.1

A contractor was blasting for a highway cut. Three-inch blastholes with six-foot burdens were drilled in sandstone rock and were loaded with ammonium nitrate and fuel oil (ANFO). The operator decided to increase his blasthole size to five inches while still using ANFO as the explosive. By substituting the numbers into Eq. (7.1), the new burden needed on the five-inch charge size can be determined. Given information:

$$B_1 = 6 \text{ ft}, De_1 = 3 \text{ in.}, De_2 = 5 \text{ in.}, \text{ and } B_2 = ?$$
$$B_2 = B_1 De_2/De_1$$
$$B_2 = 6 \times 5/3$$
$$B_2 = 30/3$$
$$B_2 = 10 \text{ ft}$$

Equation (7.1) has severe limitations, however, since it can be used only if the explosives and rock characteristics remain unchanged.

Burden adjustments for rock and explosive type. In the blasting industry, burden adjustments for rock and explosive type has always been an important necessity to be able to approximate burden dimensions at a new site, where there has been no previous blasting experience. Many equations have evolved over the years to predict the burden, based on the characteristics of both the explosive to be used and the rock mass to be shot. Allsman (1960) and Speath (1960) proposed an equation based on the following parameters:

$$B = (K \, De/12) \, (Pe/S_t)^{0.5} \tag{7.2}$$

where

$$K = \text{constant} = 0.8 \text{ for most rock}$$
$$De = \text{charge diameter}$$
$$Pe = \text{explosion pressure}$$
$$S_t = \text{rock tensile strength}$$

The difficulty in using this equation was that the borehole pressure and the tensile strength of the rock mass were not readily available. It is significant to note that the equation also indicated that burden was linear with charge diameter.

Ash (1963) published a burden equation for use in surface blasting which combined some of the variables into adjustable constants.

$$B = K_b \, De/12 \tag{7.3}$$

where

$$B = \text{burden}$$
$$K_b = \text{burden constant}$$
$$De = \text{diameter of explosive}$$

The equation was simple to use and it related burden to some constant number multiplied times the charge diameter. The article defined ranges of constants to be used based on explosive and rock types. The difference between the Ash approach and the Allsman and Speath approach was that Ash integrated the borehole pressure with the tensile strength relationship, which Allsman used as a separate function, into the first constant. In this manner, the blaster did not have to predetermine these difficult to obtain values, in fact, they could be estimated.

Ash (1968) proposed a method to adjust the constant K_b in the burden calculation by using the velocity of the explosive squared times the density of the explosive. This method of adjustment seemed to work reasonably well in midrange; however, at both ends of the velocity range, the compensations in burden were extreme.

Konya (1972) proposed a burden equation similar to the Ash equation.

$$B = 3.15 \, De \, (SGe/SGr)^{0.33} \tag{7.4}$$

where

B = burden in ft
De = diameter of explosive in in.
SGe = specific gravity of explosive
SGr = specific gravity of rock

In this equation the constant was defined as 3.15 and the constant could be adjusted by using a ratio of the specific gravity of the explosive divided by the specific gravity of the rock, both raised to the one-third power. This approach gave near identical values in the midrange as the Ash equation, however, both ends of the range were more accurately defined.

The adjustments to the constants made by Konya function well as a first approximation for both explosives and rock types worldwide; however, field personnel found it difficult to work with a relationship which contained a quantity raised to the 0.33 power.

Konya (1983) proposed another burden equation, which would give similar results to those using the earlier burden formula. This new equation was simple to use, required no power functions and was ideally suited for field use.

To easily approximate burden, the following empirical formula is helpful:

$$B = [(2SGe/SGr + 1.5)] \, De \tag{7.5}$$

where

B = burden in ft
SGe = specific gravity of explosive
SGr = specific gravity of rock
De = diameter of explosive in in.

Example 7.2

An operator has designed a blasting pattern in a limestone formation using six-inch blastholes. The six-inch blastholes will be loaded with a bulk slurry. The slurry has a specific gravity of 1.3. Limestone has a specific gravity of 2.6. Equation (7.5) can be used to determine the burden (Rock Density given in Table 7.1).

$$B = [(2 \, SGe/SGr) + 1.5] \, De$$
$$B = [(2 \times 1.3/2.6) + 1.5] \, 6$$
$$B = 15 \text{ ft}$$

TABLE 7.1 Rock Densities

Rock Type	Specific Gravity	Ton/Yd³ `
Basalt	2.8–3.0	2.57
Diabase	2.6–3.0	2.36
Diorite	2.8–3.0	2.50
Dolomite	2.8–2.9	2.43
Gneiss	2.6–2.9	2.43
Granite	2.6–2.9	2.30
Hematite	4.5–5.3	4.12
Limestone	2.4–2.9	2.23
Marble	2.1–2.9	2.09
Micaschist	2.5–2.9	2.30
Quartzite	2.0–2.8	2.16
Sandstone	2.0–2.8	2.03
Shale	2.4–2.8	2.16
Slate	2.5ᴸ–2.8	2.23
Trap Rock	2.6–3.0	2.36

In the general case, burdens which are used on the job will be reasonable if they are within plus or minus 10% of the value obtained from Eq. (7.5). Rock density is used in Eq. (7.5) as an indication of rock strength. There is a mathematical relationship between rock density and rock strength. The denser the rock, the more energy needed to overcome its tensile strength for breakage to occur. There is also a relationship between the amount of energy needed to move rock. The denser the rock the more energy is needed to move it. Explosive strength characteristics can be approximated using specific gravity because, in general, the denser explosives have more energy per unit volume. If the strength of explosives were the same on a unit weight basis, then strength would be proportional to the density. However, there are differences also in explosive energy on a unit weight basis. Those differences as compared to the differences in density are normally quite small, which allow the use of Eq. (7.5) as a first approximation.

The general equations proposed for burden selection used the specific gravity of the explosives as an indicator of energy. The new generation of slurries which are called emulsions have somewhat different energies but near constant specific gravity. The burden equations thus far proposed will define a reasonable burden but will not differentiate between the energy levels of some explosives such as emulsions. In order to more closely approach the proper burden for a test shot, we can use an equation that uses relative explosive energy rather than explosive specific gravity. Relative energy is the energy level as compared to a standard explosive. The standard explosive is defined as ammonium nitrate and fuel oil which is defined to have an energy level of 100. To use the energy equation, we would consider the relative bulk strength (relative volume strength) of the explosive. We have found that the relative bulk strength values which result from data obtained from bubble energy tests, normally produce reasonable results. Working with relative energies, however, can be somewhat misleading since relative energies can be calculated by other methods instead of bubble energy test data. The explosive in the borehole environment may not be as efficient as would have been expected from the underwater test data. The equation that uses relative energy is

$$B = 0.67 \ De \ (St_v/SGr)^{0.33} \qquad\qquad (7.6)$$

where

B = burden in ft
De = diameter of explosive in in.
St_v = relative bulk strength (ANFO = 100)
SGr = specific gravity of the rock

Corrections for numbers of rows.   Many blasting operations are conducted using one or two rows of blastholes. In this case the burden between the first row and the face and the burden between the first and second row would be equal. On some blasts, however, three or more rows of blastholes are used. When blasthole timing is not correct, it is more difficult to break the last rows of holes in multirow blasts because the previous rows are adding additional resistance and added confinement on the later rows. This commonly also occurs in buffer blasting. *Buffer blasting* is blasting to a face where the previously shot rock has not been removed. To adjust the burdens in the third, fourth, and subsequent rows we can use the correction factor, Kr, as shown below. The burden for our test shot would, therefore, be the originally calculated burden times Kr.

Corrections for Number of Rows	Kr
One or two rows of holes	1.00
Third and subsequent rows or buffer blasts	.90

Geologic correction factors.   No one number will suffice as the exact burden in a particular rock-type because of the variable nature of geology. Even when strength characteristics are unchanged the mode of rock deposition and geologic structure must also be considered in the blast design. The manner in which the beds are dipping either influences the amount of explosive used or influences the design of the burden in the pattern.

There are two rock strengths that the explosive energy must overcome. There is the tensile strength of the rock matrix and the tensile strength of the rock mass. The tensile strength of the matrix is that strength which we can measure using the brazilian or modulus of rupture test conducted on a uniaxial testing machine. Mechanical testing procedures would dictate that a massive undamaged sample of material be used for testing. We bias our test results because we use intact samples rather than those that are already broken. By doing so, we are measuring only the matrix strength and not the strength of the rock mass. The mass strength can be very weak while the matrix strength can be strong. For example, we can have a very strong rock that is highly fractured, broken, foliated, and laminated. The rock bank, however, could be on the verge of collapse simply due to the rock structure.

To approximate the deviation from the normal burden formula and average rock conditions which results from unusual rock structure, we can incorporate into our formula two constants, Kd which is a correction for the rock deposition and Ks, the correction for the geologic structure. Kd values range from 1.0 to 1.18 and describe the dipping of the beds. The classification method is broken into three general cases of deposition: beds steeply dipping into the pit or cut, beds steeply dipping into the face or into the massive rock, and other cases of deposition.

Corrections for Rock Deposition	Kd
Bedding steeply dipping into cut	1.18
Bedding steeply dipping into face	0.95
Other cases of deposition	1.00

The correction for the geologic structure will take into account the fractured nature of the rock in place, the joint strength and frequency as well as cementation between layers of rock. The correction factors for structure ranges from 0.95 to 1.30. Massive intact rock would have a Ks value of 0.95 while heavily broken fractured rock could have a Ks value of about 1.3.

Corrections for Geologic Structure	Ks
Heavily cracked, frequent weak joints, weakly cemented layers	1.30
Thin well-cemented layers with tight joints	1.10
Massive intact rock	0.95

To better understand the use of the correction factors, let us look at Ex. 7.3.

Example 7.3

The rock formation is a horizontally bedded limestone (specific gravity = 2.6) with many sets of weak joints. It is highly laminated with many weakly cemented beds. The explosive will be a cartridged slurry (relative bulk strength of 130) with a specific gravity of 1.2. The five-inch diameter cartridges will be loaded into 6.5 in. diameter wet blastholes.

$$B = 0.67 \, De \, (St_v/SGr)^{0.33}$$

$$B = 0.67 \, (5) \, (130/2.6)^{0.33}$$

$$B = 12.3 \text{ ft}$$

Correction for Geologic Conditions

$$B = Kd \times Ks \times B$$

$$B = 1 \times 1.3 \times 12.3$$

$$B = 16 \text{ ft}$$

We first calculated the average burden using Eq. (7.6). With five-inch diameter cartridges, the average burden is 12.3 feet. When the geologic correction factors are applied, we would have a burden of 16 feet.

Stemming Distance

Stemming distance refers to the top portion of the blasthole normally filled with inert material to confine the explosive gases. In order that a high-explosive charge functions properly and releases the maximum energy, the charge must be confined in the borehole. Adequate confinement is also necessary to control airblast and flyrock. A mathematical relationship for stemming is

$$T = 0.45 \; De \; (St_v/SGr)^{0.33} \tag{7.7}$$

where

$$T = \text{stemming in ft}$$
$$De = \text{diameter of the explosive in in.}$$
$$St_v = \text{relative bulk strength}$$
$$SGr = \text{specific gravity of rock}$$

However, since this equation uses the same terms as the burden equation, the only difference is the constant, which changes from 0.67 to 0.45, we can, therefore, simplify the stemming equation and relate it directly to the burden.

$$T = 0.7B \tag{7.8}$$

where:

$$T = \text{stemming in ft.}$$
$$B = \text{burden in ft}$$

In most cases, a stemming distance of $0.7 \times$ burden is adequate to keep material from ejecting prematurely from the hole. It must be remembered that stemming distance is proportional to the burden, therefore, charge diameter, specific gravity, or strength of explosive and specific gravity of rock were all needed to determine the burden. The stemming distance is also a function of these variables. If the blast is poorly designed, a stemming distance equal to $0.7 \times B$ may not be adequate to keep the stemming from blowing out. In fact, under conditions of poor design doubling, tripling, and quadrupling the stemming distance may not ensure that holes will function properly, therefore, the average stemming distance previously discussed is only valid if we assume that the shot is functioning properly.

Example 7.4

In Ex. 7.2, a six-inch diameter blasthole was used in limestone. It was determined that a 15-foot burden would be a good first approximation. To determine the stemming distance needed in that blast, we would do as follows:

$$T = 0.7 \times B \quad \text{(for crushed stone or drilling chips)}$$
$$T = .7 \times 15 \text{ ft}$$
$$T = 10.5 \text{ ft}$$

where

$$T = \text{stemming in ft}$$
$$B = \text{burden in ft}$$

The common material used for stemming is drill cuttings, since they are conveniently located at the collar of the blasthole. However, very fine cuttings commonly called drilling dust make poor stemming material. If you use drill cuttings heavy with drilling dust, approximately 30% or $0.3 \times B$ more stemming would have to be used than if the crushed stone were used for the stemming material. In instances where solid rock is located near the surface of the bench (cap rock), operators often bring the main explosive column as high as possible to break this massive zone. However, they do not want to risk the possibility of blow-out, flyrock, and airblast. In cases such as this, it is common to bring crushed stone to the job site to use as stemming material. In Ex. 7.4, where the stemming distance was calculated, if drilling dust were used instead of crushed stone or drilling chips, it may be necessary to increase the stemming depth to equal burden distance. Drilling dust makes poor stemming material, since it will not lock into borehole walls and is easily ejected.

If stemming distances are excessive, poor top breakage will result and the amount of backbreak will increase. When a blast functions properly, the stemming zone will gently lift and slowly drop onto the broken rock pile after the burden has moved out. This action is illustrated in Fig. 7.2.

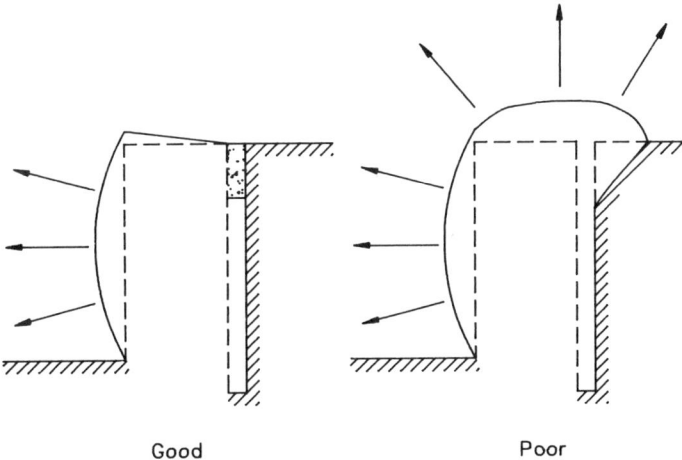

Good   Poor

Figure 7.2 Behavior of stemming zone.

Optimum size of stemming material.   Selection of the proper size of stemming material is important if one wants to minimize the stemming depth in order to break cap rock. Very fine drilling dust will not hold into the blasthole. Very coarse materials have the tendency to bridge the hole when loading and may be ejected. The optimum size of stemming material would be material that has an average diameter of approximately 0.05 times the diameter of the blasthole. Material must be angular to function properly. The best size would be determined as follows:

$$SZ = 0.05 \, Dh \tag{7.9}$$

where

SZ = particle size in in.
Dh = blasthole diameter in in.

River gravel of this size, which has become rounded, will not function as well as crushed stone. Upon detonation of the explosive in the blasthole, stemming particles will be compressed to mortar consistency for a short distance above the charge (Fig. 7.3).

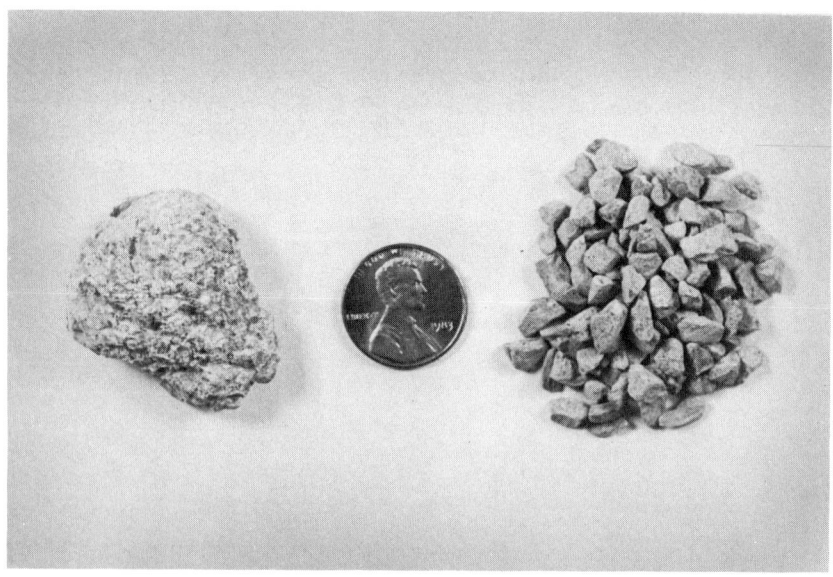

Figure 7.3 Stemming material compaction immediately above charge. Compact material results from crushed stone (on left).

Subdrilling

Subdrilling is a term that defines the depth to which a blasthole will be drilled below the proposed grade to ensure that breakage will occur to the grade line. Blastholes normally do not break to full depth. On most construction or mining projects, other than coal surface mining, subdrilling is used unless, by coincidence, there is either a soft seam or a bedding plane located at the grade line. If this occurs, no subdrilling would be used. In fact, blastholes may be backfilled a distance of 6 to 12 charge diameters to confine the gasses and keep them away from a soft seam (Fig. 7.4).

On the other hand, if there is a soft seam located a short distance above the grade line and below that there is massive material, it is not uncommon to have to subdrill considerably deeper in order to break the material below the soft seam. As an example, Fig. 7.5 indicates a soft seam one foot above the grade. In this case, a subdrilling approximately equal to the burden distance was required below the grade to ensure breakage to grade.

In most instances, subdrilling is approximated as follows:

$$J = 0.3\,B \qquad\qquad (7.10)$$

where

$$J = \text{subdrilling in ft}$$
$$B = \text{burden in ft}$$

The subdrilling must not contain drill cuttings, mud, or any rock materials. If borehole walls crumble and naturally fill in, drilling must be deeper than the subdrilling previously discussed so that at the time of loading the calculated amount of subdrilling is open and will contain explosives.

In order to get a flat floor in an excavation, it makes good economic sense to drill to a depth below grade, which ensures, in spite of random drilling depth errors

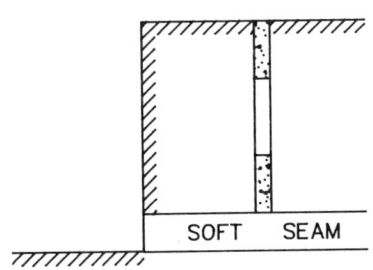

Figure 7.4 Borehole backfilled above soft seam.

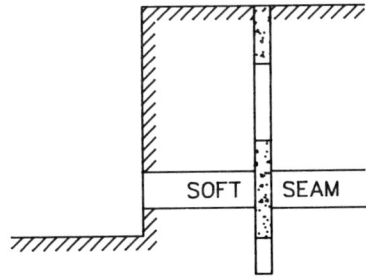

Figure 7.5 Stemming across soft seam off bottom.

and caving holes, that all hole bottoms will be down to the proper depth at the time of loading. If drilling is done slightly deeper than required and some holes are too deep at the time of loading, the blaster can always place drill cuttings in the bottom of those holes to bring them up to the desired height. The blaster, however, does not have the ability, at the time of loading, to remove excessive cuttings or material which has fallen into the hole.

The maximum tensile stress produced in the burden as a result of subdrilling is illustrated in Fig. 7.6. The zone which is cross-hatched indicates the zone of maximum tension in the rock. You can see that the zone of maximum tension in Fig. 7.6(a) is located further from the bottom of the floor level than in Fig. 7.6(b). In Fig. 7.6(b), where subdrilling was used, there is a larger zone of maximum tension and it occurs closer to floor level or the zone which must be sheared.

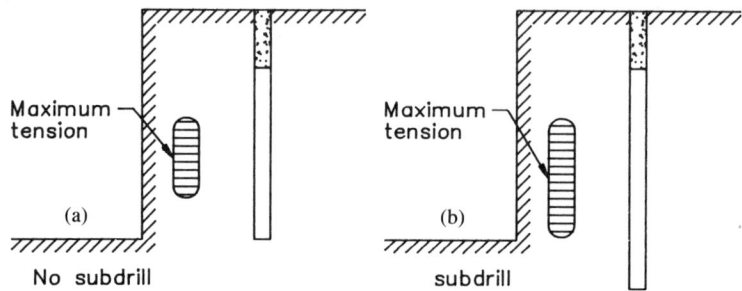

Figure 7.6   Zones of maximum tensile stress.

Example 7.5

A six-inch blasthole was used in Ex. 7.2 in limestone rock. The burden was determined to be 15 feet. The amount of additional drilling or subdrilling which would be needed below grade to ensure breakage to grade is determined by using Eq. (7.10).

Solution

$$J = 0.3 \times B$$
$$J = 0.3 \times 15$$
$$J = 4.5 \text{ ft}$$

Selection of Blasthole Size and Bench Height

The selection of the proper size blasthole for any job requires a two-part evaluation. The first part would consider the effect of the drillhole size on fragmentation, airblast, flyrock, and ground vibration. The second would consider drilling economics.

The effect of the design on fragmentation, airblast, flyrock, and ground vibration would all have to be assessed. In general, the larger the hole size, the more problems are possible with airblast, flyrock, ground vibration, and fragmentation.

To gain insight into the potential problems that can result requires the consideration of the stiffness ratio, which is the bench height divided by the burden distance, or L/B. Table 7.2 is a summary of general potential problems as related to the stiffness ratio.

With the help of Table 7.2, the operator can determine the potential for the unwanted effects which were previously discussed, and determine how much of a trade-off to make with the drilling and loading economics and these factors. The more massive the rock in a production blast, the more probable the outcome listed in Table 7.2.

TABLE 7.2   Potential Problems as Related to Stiffness Ratio (L/B)

Stiffness Ratio	Fragmentation	Airblasts	Flyrock	Ground Vibration	Comments
1	poor	severe	severe	severe	Severe backbreak toe problems. Do not shoot. Re-design.
2	fair	fair	fair	fair	Redesign if possible.
3	good	good	good	good	Good control and fragmentation.
4	excellent	excellent	excellent	excellent	No increased benefit by increasing stiffness ratio above 4.

Example 7.6

The Ajax Construction Company is removing a cut for a highway project. The maximum bench height is 30 ft deep. Because of the small loading equipment, fragmentation must be good. The operator has track drills capable of drilling up to five-inch diameters and a rotary drill capable of going up to 7–7/8 diameter in his equipment inventory. What hole size should be selected based on the local conditions?

Solution   Questions which must be answered.

a. Are the blastholes wet? Should cartridged or bulk powder be used? (Assume dry holes: ANFO used as explosive.)
b. What amount of explosive can be loaded per blasthole or per deck without having vibration problems?
c. Must airblast and flyrock be totally avoided and should blasting mats be used?

Since fragmentation must be good, select an L/B ratio of 3. The explosive selected based on the answer to question "a" has a density of 0.8 and the rock density is 2.6. Equation (7.5) can be used and solved for D_e.

$$\text{If } L/B = 3 \text{ and } L = 30 \text{ ft}$$

then

$$B = L/3$$
$$B = 30/3$$
$$B = 10 \text{ ft}$$

Using Eq. 7.5

$$B = [(2 \ SGe/SGr) + 1.5] \ De$$

Substituting 10 for B and rearranging the equation gives

$$De = 10/[(2 \ SGe/SGr) + 1.5]$$
$$De = 10/[(1.6/2.6) + 1.5]$$
$$De = 4.72 \text{ in.}$$

The number found with these calculations would not necessarily be the optimum hole size. It would be the maximum hole size one would want to use to minimize the conditions previously discussed. The vibration limitations, if any, would now be checked using the propagation equation or scaled distance value discussed in Chap. 3. Any charge diameter larger than 4.72 in. would increase the probability of coarse fragmentation, airblast, flyrock, and ground vibration per lb of explosive used.

A simple method used to approximate a blasthole length where the stiffness ratio is above two is given in Fig. 7.7 and is often called the "Rule of Five."

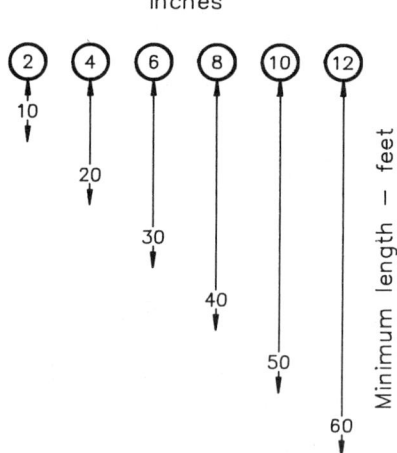

Figure 7.7   Blasthole minimum length for stiffness ratio of two.

$$L = 5 \times De \qquad\qquad (7.11)$$

where

$$L \ = \ \text{minimum bench height in ft}$$
$$De = \text{diameter of explosive in in.}$$

The minimum length of blasthole in feet is approximated by multiplying the charge diameter in inches by five.

Primer Number Determination

The considerations for the design of a single borehole cannot be complete until we check whether the number of primers in the blasthole will fit with the final design.

We can mathematically determine the maximum column length which can be efficiently detonated by a single primer. In order to do this we must have information on rock characteristics and must assume that the explosive is firing at the rated detonation velocity.

In order to make the analysis meaningful we must assume that the rock mass is uniform without any mud or soft seams in the formation. We must also assume that the rock will crack at the average speed of about 20% of its longitudinal wave velocity. We will also conservatively assume that the first major loss of energy could only occur when the radial cracks first reach the open face. We will assume that radial cracks will also travel from the primer which for our example will be located at floor or grade level. To help in this analysis we will use the symbols and borehole geometry as shown previously in Fig. 7.1, which shows a cross section of a blasthole.

Approximately 30 millionths of a second, after the primer detonates, the first cracks begin to form on the borehole walls at the primer. We can determine the time it would take for a fracture at floor level to move through the burden and reach the open face. That time t_1 can be calculated from Eq. (7.12).

$$t_1 = B/Vc \qquad\qquad (7.12)$$

where

$$t_1 = \text{time in ms}$$
$$B = \text{burden in ft}$$
$$Vc = \text{velocity of crack in ft/sec}$$

Note: Minimum average crack speed $Vc = 0.2\ Vp$. The longitudinal wave velocity is designated as Vp.

During the time it takes the radial cracks to travel from the primer to the face, we want full detonation of the entire explosive column. Since the primer is at floor level, the entire powder column length is not considered in the calculation, only the length of powder column between the primer and the stemming. The primer is assumed to be bi-directional, therefore, it will take more time to shoot from floor level up to the stemming than from the floor level to the bottom of the subdrill which is normally a much smaller distance. We can set up a general equation for the time it takes to detonate the column of explosives from the primer to the bottom of the stemming. It is shown as t_2 and can be calculated from Eq. (7.13).

$$t_2 = (PC - J)/Ve \qquad\qquad (7.13)$$

where

$$t_2 = \text{time in ms}$$
$$PC = \text{powder column length in ft}$$
$$J = \text{subdrill depth in ft}$$
$$Ve = \text{detonation velocity in ft/sec}$$

To determine the maximum length of the powder column we set $t_1 = t_2$ and solve for the powder column length PC shown in Eq. (7.14).

$$t_1 = t_2 \qquad\qquad (7.14)$$
$$B/Vc = (PC - J)/Ve$$
$$PC = B\ Ve/Vc + J$$

Substituting

$$Vc = 0.2\ Vp$$
$$PC = 5B\ Ve/Vp + J$$

The powder column length which could effectively detonate from one primer located at floor level is given by Eq. (7.13). Let us look at an example and see whether or not the addition of the second primer from a technical standpoint would be justified.

Example 7.7

We will use as an example a four-inch diameter blasthole loaded with ammonium nitrate and fuel oil. The explosive will have a detonation velocity of 9000 ft per sec. The rock will have a longitudinal wave velocity (Vp) of 20,000 ft per sec. The blast plan calls for a ten-foot burden, 7 ft stemming and 3 ft subdrilling. In the design it is planned to have a powder column length that would be 40 ft long. We want to determine whether adding a second primer on the same delay period would be justified.

$$PC = 5B\ Ve/Vp + J$$
$$PC = [5(10)(9000)/20000] + 3$$
$$PC = 25.5$$

Substituting the actual data into Eq. (7.14) shows that a single primer at floor level could effectively cover 25.5 ft of explosive column charge before the first cracks would reach the open face. This leaves 14.5 ft of explosive column that may not have reacted by the time the first crack reached the face with a 40 ft powder column.

If you were to assume that the second primer would go off at exactly the same time as the first then the addition of a second primer near the top of the powder column would be justified. To reach a realistic solution, however, we must take into account the tolerance in the firing times of the caps. If you used a high-precision cap of low-period number, you may have a very small difference in firing time of the two caps of the same period. On the other hand, if you used a long period delay such as a 400 ms or 500 ms delay, there is a high probability that only one of the primers will fire since the cap in the second primer will most probably be initiated by the powder column itself.

To understand why this occurs, let us look at the actual time in milliseconds it would take for a 25.5 ft powder column to react. If we take the 25.5 ft and subtract from it the 3 ft of subdrill we would get 22.5 ft of explosive. We can use Eq. (7.13) to determine how long it could take for this explosive to react. Divide the 22.5 ft of explosive by the 9000 ft per sec reaction velocity and we find that the total reaction time would be only 2.5 ms.

If we conservatively assume that high-precision caps would have a tolerance in firing time of plus or minus 1% of the firing period, we can see why the addition of a second primer firing on the same time period is sometimes beneficial and sometimes it is not. If, for example, the firing time of the cap in the hole was 25 ms, then a 1% tolerance on 25 ms would be .25 ms. If we assume that one cap can fire fast and the other cap fires slow we would add the plus and minus tolerances and get 0.5 ms, which is the possible deviation in firing time between the two primers. In this case, the use of the second primer could be justified.

Let us next consider using a 400 ms cap delay in each of the primers and determine whether or not the results would be productive. If we make the same assumptions on cap accuracy, a 400 ms cap could have a plus or minus 4 ms tolerance in firing time or in the worst case a total of 8 ms. We have previously stated that 22.5 ft of the powder column, would shoot in 2.5 ms. We unfortunately have the possibility of an 8 ms difference in firing time of the caps which is much greater than the firing time of the entire explosive column. It should now be obvious why the double priming technique sometimes works and sometimes doesn't with higher period caps.

If the caps fire at near the same time, then there is a beneficial effect by using the second primer, on the other hand, if the cap fires on the outer ranges of their tolerance, the second primer is never fired by the cap because the powder column is totally consumed in only 2.5 ms.

Because of the random deviation in firing times of the cap, each and every hole will not behave the same. We have made the assumption that the bottom primer located near floor level will fire first and that the second primer will fire sometime later, however, this is not a realistic assumption. The top primer could fire first and the second primer would never fire since the powder column would be consumed before the cap could initiate. This procedure can lead to random results since different effects occur if blastholes are collar primed or bottom primed. The blaster would have no control whatsoever on which cap would ever fire first.

Drilling Accuracy

The accurate execution of any blasting plan involves accurate blasthole drilling. It is not uncommon to find blasthole on the surface drilled up to 50% in error on burdens and spacings. These large errors change the energy distribution or energy density in the rock mass and cause boulders, flyrock, high-ground vibration, as well as other problems associated with blasting. Best results occur if blastholes are drilled within one-hole diameter of their desired location. Commonly, a driller will have problems collaring a hole at a specific location and will have to move over. That type of occasional drilling error is acceptable, however, it should not occur on the majority of the holes within the blasting pattern.

The inclination of the mast on the drill is an important consideration, especially in deep holes. Mechanical and electrical devices are available which help the driller position the mast so that each hole enters the ground at the same angle. Visual estimation of the drill angle is a poor procedure. Figure 7.8 shows the deviation of the drill at the bottom of the hole for different degrees of error in alignment, for example, it can be seen that on a 50-ft bench a 5-degree error in

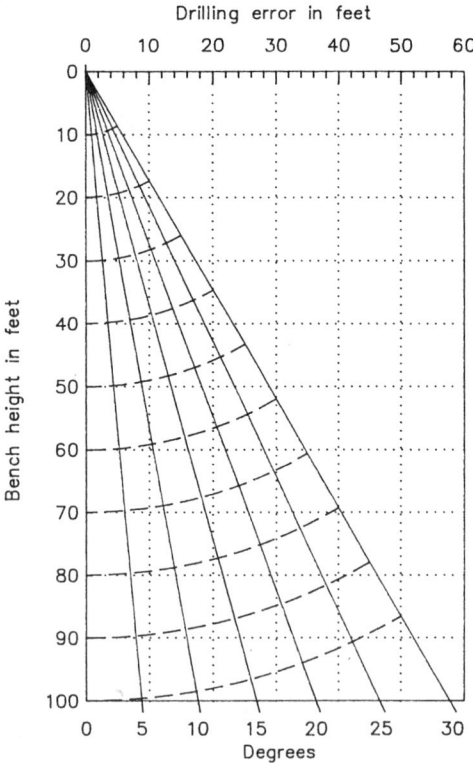

Figure 7.8 Drilling error which results from drill deviation.

alignment, will result in a 4 ft difference in burden at the bottom of the hole. If the error in drilling was 10 degrees then the error in burden would be about 9 ft. The blasthole will also be short and the subdrilling will not be deep enough. In order to properly break at the floor level the blasthole must be in the proper location, both at the collar and bottoms of the holes to provide the proper energy density within the rock mass.

Angle Drilling

Different types of drills have different angling capabilities. In general, production holes are vertically drilled in the United States and angle drilling normally is only used for controlled blasting applications such as presplitting or trim blast. In many foreign countries, however, angle drilling is used on production holes as well as controlled blasting applications. To better understand the rationale for angle versus vertical drilling, you would have to compare the advantages and disadvantages of both methods and look at them from a site specific application (Fig. 7.9).

ADVANTAGES OF ANGLE DRILLING

1. less backbreak
2. less problems at grade
3. more throw, especially on low benches
4. better fragmentation on low benches
5. loose rock better held on face by gravity

DISADVANTAGES OF ANGLE DRILLING

1. harder to collar holes
2. difficult to maintain accurate angle
3. more problems with geologic discontinuities
4. easier to hang steel in holes

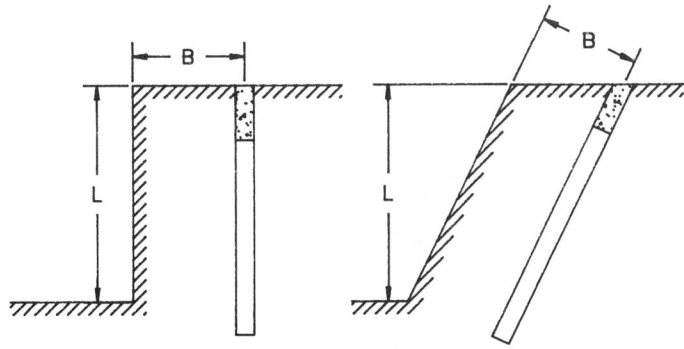

Figure 7.9 Vertical versus angle drilling.

5. more difficult to load explosives

6. often impossible with drilling machines being used

ROCK FRAGMENTATION

Two important principles must be correctly applied to control fragmentation size. The proper amount of energy must be applied at strategic locations within the rock mass. The energy must also be released at a precise time to allow the proper interactions to occur.

The energy distribution within the rock mass is further broken down into two distinct areas. First we must have sufficient energy to break the rock mass, by using the proper amount of explosives. The explosive must also be placed in a geometric configuration, whereby, the energy is maximized for fragmentation. This geometric configuration is commonly called the blasting pattern.

The release of the energy at the wrong time can change the end result, even though the proper amount of energy is strategically placed throughout the rock mass in the proper pattern. If the initiation timing is not correct, differences in breakage, vibration, airblast, flyrock, and backbreak can occur. This discussion does not consider the effects of the timing on the release of the energy, we will only consider for the purposes of this section, the strategic placement of the proper amount of energy in a correct blasting pattern.

The study of the concerns of fragmentation go back to the early days of blasting. Blasters realized that on some blasts the energy was very efficiently used in the breakage process, and on others, very little energy was used in an efficient manner and instead a great deal of noise, ground vibration, airblast, and flyrock resulted with little breakage. There have been many empirical methods that have surfaced over the decades, suggesting methods of design which would more efficiently utilize this energy. These design methods would also give the blaster a way of producing consistency in his results, by applying similar techniques under different circumstances and in different rock masses.

Kuznetsov (1973) did research on fragmentation. His work relates a mean fragmentation size to the powder factor of TNT and to the geologic structure. Kuznetsov's work was very important, since it showed that there was a relationship between average fragmentation size and the amount of explosive used in a particular rock type. Although mean fragmentation size could be predicted, it told nothing about the amount of fines produced or the amount of boulders. The same mean size could result if we have four-foot diameter boulders and dust, or if we had every piece of breakage exactly two feet in size. What was then needed was a way of determining the actual size distribution, not just the mean size. The actual size distribution is a function of the pattern, the manner in which the explosive is geometrically applied to the rock mass.

Kuznetsov equation. The original Kuznetsov equation is given as

$$\bar{x} = A(V_o/Q)^{0.8}Q^{0.167} \tag{7.15}$$

where

$\bar{x}$ = mean fragment size, cm

A = rock factor =   1 for extremely weak rock
                                7 for medium rocks
                                10 for hard, highly fissured rocks
                                13 for hard, weakly fissured rocks.

V_o = rock volume (cubic meters) broken per blasthole, taken as burden $\times$ spacing $\times$ bench height

Q = mass (kg) of TNT which is equivalent in energy to that of the explosive charge in each blasthole

Normally the explosive in the subdrill is excluded. Although necessary, it seldom contributes significantly to fragmentation.

With the use of the original Kuznetsov equation and the modifications supplied by Cunningham (1983), you can determine the mean fragmentation size with any explosive and the index of uniformity. With this information, a Rosin Rammler projection of size distribution can be made.

Size distribution. Cunningham realized that the Rosin Rammler Curve had been generally recognized as a reasonable description of fragmentation for both crushed or blasted rock. One point on that curve, the mean size, could be determined by using the Kuznetsov equation. To properly define the Rosin Rammler Curve, what was needed was the exponent "n" in the following equation:

$$R = e^{-(x/x_c)^n} \tag{7.16}$$

where

R = proportion of material retained on screen

x = screen size

x_c = $\bar{x}/(0.693)^{1/n}$

n = index of uniformity.

To obtain this value, Cunningham used field data and regression analysis of the field parameters that were previously studied and obtained "n" in terms of

1. drilling accuracy
2. ratio of burden to blasthole diameter
3. staggered or square drilling pattern
4. spacing/burden ratio
5. ratio of charge length to bench height

The combination of the algorithms thus developed along with the Kuznetsov equation, became know as the "Kuz-Ram Model." The algorithm is

$$n = (2.2 - 14 \, B/d) \, (1 - W/B) \, [1 + (A - 1)/2] \, L/H \qquad (7.17)$$

where

d = charge diameter (*mm*)
B = burden (*m*)
W = standard deviation of drilling accuracy (*m*)
A = spacing/burden ratio
L = charge length above grade level (*m*)
H = bench height (*m*)

A further development which enables the use of different explosives other than TNT, was incorporated into the Kuznetsov equation by Cunningham. The final equation to determine average fragmentation size is shown below

$$\bar{x} = A \, (V/Q)^{0.8} \, Q^{0.17} \, (E/115)^{-0.63} \qquad (7.18)$$

The "E" is a Relative Weight Strength term of the actual explosive (where ANFO = 100) while the Relative Weight Strength of TNT is 115. The strength values are available from the explosives manufacturers' product data sheets.

Field results.   Kuznetsov's initial studies were done in models of different materials, and later applied to surface mining operations. There was some difference between the fragmentation measured and the predictions, as was to be expected, considering the nature of mining and the variability of rock. One would expect correlation to be best in the model work where the materials properties can be tightly controlled. The larger the scale of the operation and the bigger the holes and more varied the rock, the greater would be the expected deviation between the predicted results and the measured results in fragmentation. The actual measurement of fragmentation from large scaled blasts is extremely difficult and as a result there are only a few such measurements in existence, some of these were done with photographic techniques. The biggest problem with photographic assessment would be in the fines content.

Verification of the equations were done both in small scale blasting and also on large scale blasts. The US Bureau of Mines (1973) conducted tests on small scale blasts in limestone. Fragments from the blast were collected and sized. This data was put into the model for evaluation. It is interesting to note that the data fit the predictions quite reasonably with less than a 10% variation from the measured range over the greatest part of the curve. A typical result from both the predicted and the measured data is shown in Fig. 7.10. The Kuz-Ram method was used to evaluate some of the fragmentation from overburden blasts at Australian coal mines. The model gave an extremely good correlation in the coarse section of the curve, but indicated rather more fines than are given by the photographic analysis, which was used.

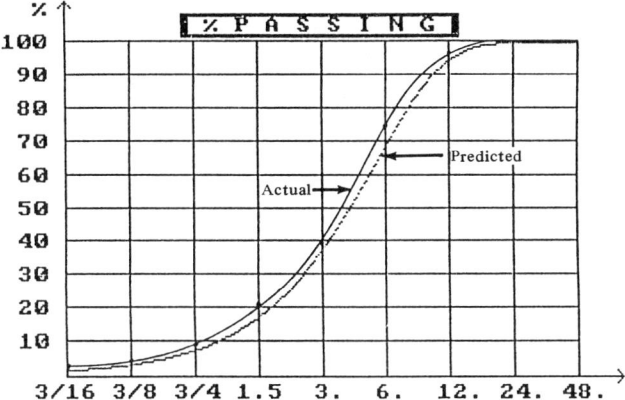

Figure 7.10 Predicted and actual fragmentation distributions.

Limitation on the Kuz-Ram Model. A simple model as this, requires caution in its use and the following factors should be understood:

a. The Spacing/Burden ratio applies only to the drilling function, not the timing, therefore, spacing is always considered along the row where burden is considered the distance between rows, which parallel the face. The layout on this blast can never be such that the spacing to burden ratio is greater than two.

b. It is assumed that reasonable timing sequences are used which will enhance or maintain fragmentation.

c. The explosive should actually yield energy close to its Relative Weight Strength for the diameters that are being used on the job.

d. Jointing and bedding, especially in the case of loose jointing, more closely spaced than the drill pattern can effect the size distributions. Maximum sizes could be controlled by geologic features rather than the explosives energy released from the blasting process.

Effects of blasting parameters on "*n*." It would be desired to have uniform fragmentation in a blast, avoiding both excessive fines and boulders. If this is to be obtained, high values of "*n*" are preferred. The value for "n" increases as

the burden/hole diameter decreases

the drilling accuracy increases

the charge length/bench height increases

the spacing/burden increases

the use of a staggered pattern rather than a square pattern.

The effects of stronger explosives. In many operations, a standard drilling pattern is used. The drill pattern selection is based on considerations such as drilling capacity, or the policy of drilling well ahead of the blasting operation. Where patterns are already drilled, improvements in breakage can be made by increasing the explosive strength.

Explosive strength is determined by density as well as strength. By increasing the density, we increase the total pounds of explosives put into the blast.

Fragmentation Effects on Wall Control

In general, it can be said that the better the breakage obtained and displacement on a row-by-row shot, the better the wall control. If insufficient energy is available to break rock properly in the burden, the added burden resistance placed against the borehole causes increased confinement and will cause more fracturing (back shatter) behind the blast. If large boulders are produced from the stemming area rather than from the burden, increased backbreak especially at the top of the bench will result, thereby causing problems with subsequent drilling of patterns and the final wall will be less stable. In general, one can conclude that the higher the "n" value the better the potential wall control. One could also conclude that the lower the mean size on a specific design, the smaller the chance of causing back shatter and excessive overbreak beyond the excavation limits. Values of "n" below 1.0 should be avoided. Values of "n" between 1.0 and 1.3 indicate potential wall damage.

The fragmentation model can therefore be used for two purposes: to determine the sizing which results from the blast, and the effect of one pattern versus another on potential problems with wall control.

"Blast Fragmentation Prediction" is a commercial software package using computer models to illustrate the use of a fragmentation program based on a modified Kuz-Ram method. The results are similar to the method previously described.

The following examples (Figs. 7.11 through 7.14) illustrate the effect of changing bench height on fragmentation and wall control. Entry #1 differs from Entry #2 by a change in bench height from 60 feet to 10 feet. The input data on the two patterns are given in Fig. 7.11 and 7.12. Figures 7.13 through Figure 7.15 show the resulting changes in fragmentation size. The 10-foot bench height results in a larger average size.

The fragmentation index (n) also drops below 1.0 for the 10-foot bench height, thereby producing a condition that is likely to cause severe wall damage.

General Timing Effects on Fragmentations

The fragmentation model previously discussed can produce approximate fragmentation distributions. Distribution can be further influenced by the selection of the actual milliseconds of time delay between holes and rows. Chapter 8 will deal with methods of scaling time to produce reasonable fragmentation.

```
ACTIVE ENTRY # 1   Calculation based on Volume Strength.


     Charge diameter [inches (decimal)] .....................  4.000
     Burden [feet] ..........................................  9.000
     Spacing [feet] .........................................  9.000
     Stemming [feet] ........................................  6.000
     Subdrill [feet] ........................................  3.000
     Bench height [feet] .................................... 60.000
     Hole depth [feet] ...................................... 63.000
     Pattern factor (1 - for square, 2 - for staggered) .....  2.000
     Number of rows .........................................  5.000
     Explosive volume strength (ANFO = 100) ................ 100.000
     Specific gravity of explosive ..........................  0.850
     Rock strength (weak-fissured = 1, strong-massive = 10)..  4.000
     Explosive charge per hole [lbs] ....................... 263.000
```

Figure 7.11 Data for pattern number 1.

```
ACTIVE ENTRY # 2   Calculation based on Volume Strength.


     Charge diameter [inches (decimal)] .....................  4.000
     Burden [feet] ..........................................  9.000
     Spacing [feet] .........................................  9.000
     Stemming [feet] ........................................  6.000
     Subdrill [feet] ........................................  3.000
     Bench height [feet] .................................... 10.000
     Hole depth [feet] ...................................... 13.000
     Pattern factor (1 - for square, 2 - for staggered) .....  2.000
     Number of rows .........................................  5.000
     Explosive volume strength (ANFO = 100) ................ 100.000
     Specific gravity of explosive ..........................  0.850
     Rock strength (weak-fissured = 1, strong-massive = 10)..  4.000
     Explosive charge per hole [lbs] ....................... 32.000
```

Figure 7.12 Data for pattern number 2.

```
Entry # 1 Calculation based on Volume Strength.
Entry # 2 Calculation based on Volume Strength.
                             s c r e e n    s i z e

          3/16    3/8    3/4    1.5    3     6     12     24     48

% Passing

Entry # 1  0.0    0.2    0.6    2.1    7.2   23.2  60.7   96.3  100.0
Entry # 2  2.2    3.9    6.8   11.5   19.3   31.3  48.3   68.5   86.8

% Fraction

Entry # 1         0.2    0.4    1.5    5.1   16.0  37.4   35.6    3.7
Entry # 2         3.9    2.9    4.8    7.8   12.0  16.9   20.2   18.3

                                       Entry # 1      Entry # 2

          Average size [inches] ...........  10.189        12.775
          Fragmentation index .............   1.820         0.809


          Press:    B - Bar graph     L - Line graph    M - Menu
          To return to this screen from the Graph, press SPACE BAR.
```

Figure 7.13 Summary of fragmentation data.

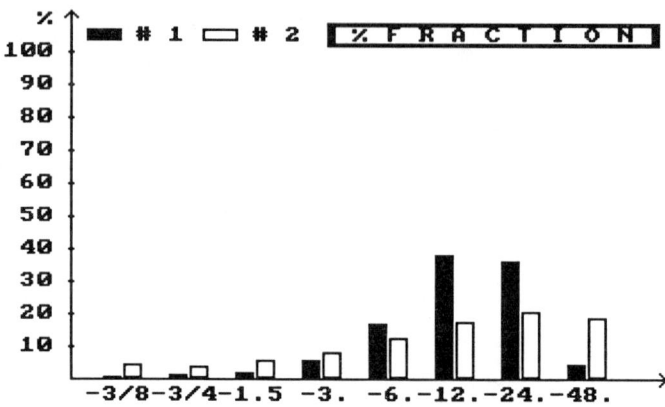

Figure 7.14   Comparison of sizes from both blasts.

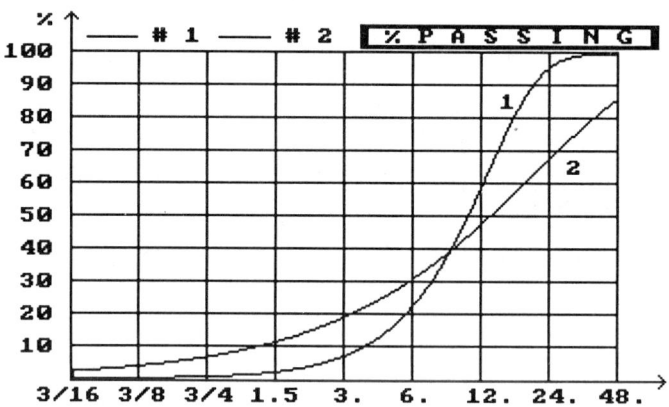

Figure 7.15   Cumulative distribution of fragmentation data.

Research conducted by Phuphaibul (1975) with small scale models, defines trends which link the effects of initiation timing to fragmentation. This research scaled the blasthole geometries with long explosive charges. Only two charges were used per test. The material broken was sized into three fractions for comparison. Different spacing ratios were used. Spacing was equal to the burden, twice the burden and three times the burden. Two different times were used. In one case the holes were shot truly simultaneously by using a single cap to initiate identical lengths of explosive column. In the second case, one hole was delayed from another. The amount of delay used for the test was scaled at three milliseconds per foot of burden.

Test data indicated that results of fragmentation sizing were duplicated with reasonable correlation achieved between identical shot geometries. The results showed that there would always be a larger percentage of large size material produced with instantaneous initiation regardless of the spacing ratio. Between spacing ratios of one and two which are commonly used in field blasting, the instantaneous initiation models produced 350 percent more large size material than the properly delayed shots. The properly delayed shots produced approximately 4% of their total volume of broken material as large pieces while the simultaneous shots produces about 14% of the total volume of large fragment (Fig. 7.16).

Deck Loading for Vibration Control

The amount of blast vibration produced is proportional to the amount of charge detonated at any one instant of time if the design geometry and timing does not change. As blasting approaches structures, you have to cut the amount of charge per delay to control vibration levels. Column loads of explosives in large boreholes may often produce more vibration than is allowable, therefore, the explosive column within the hole, is divided into sections which are fired at different instants of time. This method of separating the explosive column into different sections is called deck loading (Fig. 7.17). To efficiently break the explosive column so that one portion of the column will not detonate another, inert material is placed between the two charges. This inert material normally consists of drill cuttings. Charges can propagate across inert material if the length of inert material is insufficient. Field results indicate that wet holes are more likely to propagate than dry holes. To reduce the chance of propagation across decks, we can use the following guidelines to approximate the necessary stemming thickness between charges.

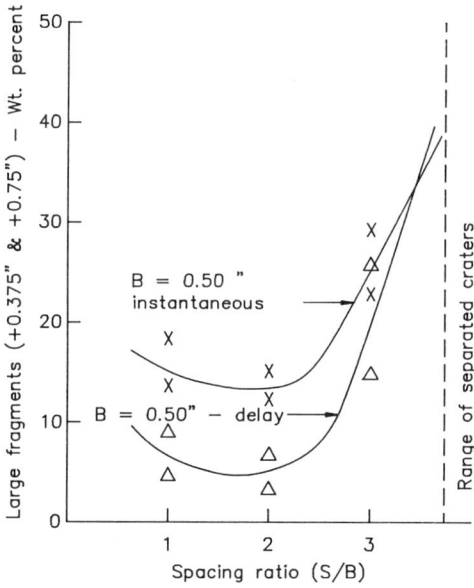

Figure 7.16 The relationship of weight percent of large fragments (+ 0.375 in. + 0.75 in.) to spacing ratio (S/B).

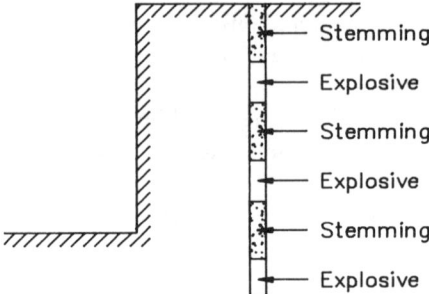

Figure 7.17   Deck loading blastholes.

For dry holes, deck stemming thickness = 6 times the hole diameter.
For wet holes, deck stemming thickness = 12 times the hole diameter.

These guidelines are effective for avoiding propagation in most cases, however, size consistency of the inert material used and the amount of water saturation
controls the amount of energy which is delivered to the adjacent deck. In some rare
instances, propagation has occurred even when these guidelines have been followed.

Blasthole Size for Different Mining Applications

Blasting economics are determined by both the drilling and blasting costs.
Large diameter holes are more economical to drill then many small diameter holes
for the same yardage produced. It is common for blasting operations, whether it be
for mining or construction purposes, to increase hole size to the maximum to reduce
drilling cost. Large diameter holes with low stiffness ratios can cause more violence
and backbreak problems. We can compare the hole diameters and blasthole depths
in Figs. 7.18 through 7.20 to get a better understanding of what is being currently
used in the noncoal sector of the mining industry. A survey was conducted in 1980
to determine blasthole diameters and hole depths used in copper mines, iron mines,
and quarries. The graph in Fig. 7.18 shows the data obtained for copper mines. The
reference line on the graphs relate the condition where L/B equals two. From the
graph, it can be seen that many copper mining operations are using stiffness ratios
that are less then two and those would be above the reference line. Figure 7.19
illustrates a similar graph for iron mines. All iron mines surveyed had stiffness
ratios less than two. If we compare the results from the copper and iron mines with
those obtained from quarries, we notice a striking difference. In quarries, most of
the data fell below the reference line indicating that most quarries used stiffness
ratios greater than two. This data may help put into proper prospective some of the

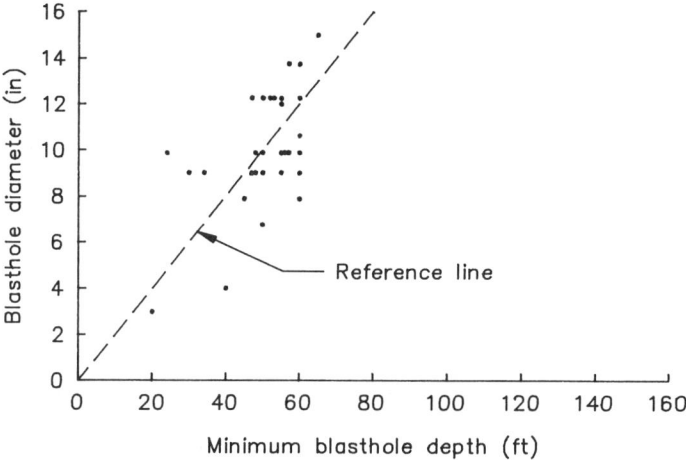

Figure 7.18 Blasthole diameter versus minimum blasthole depth for domestic copper mines. (1980)

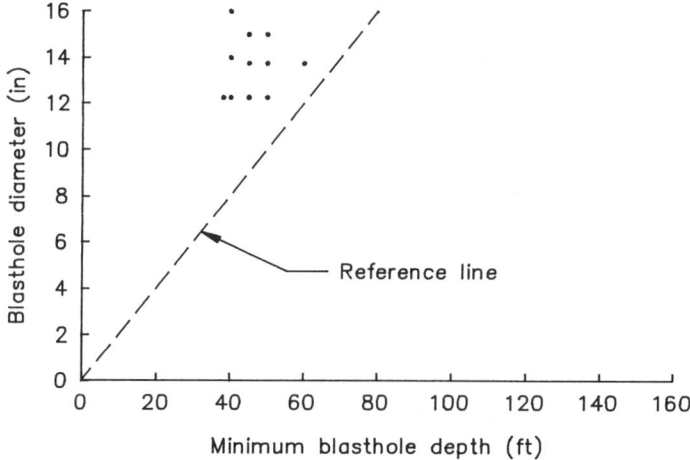

Figure 7.19 Blasthole diameter versus minimum blasthole depth for iron mines. (1980)

different effects obtained in mining operations which use large holes and short benches versus those obtained in quarries. In general, the large hole operations produce more violence and uplift while quarries produce less upward motion during blasting (Fig. 7.20). The differences in effects obtained in these operations are not due to the type of rock that is blasted. All blasting operations are laid out using the same principles and are controlled by the same laws of physics. The major differences are caused by the stiffness ratio which is directly related to the blasthole diameter, its resulting burden, and the bench height.

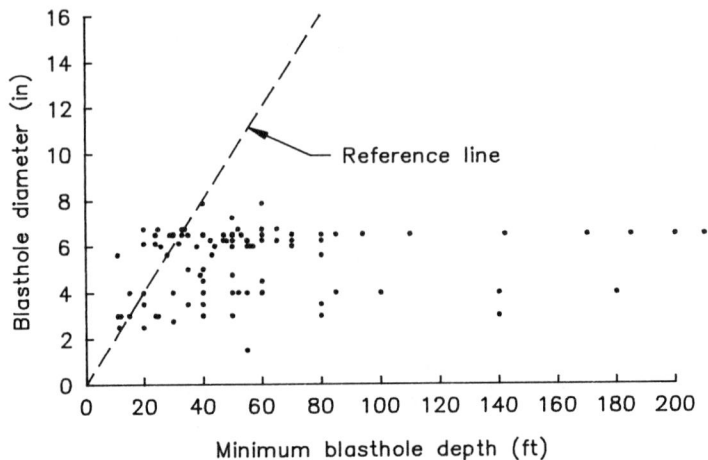

Figure 7.20   Blasthole diameter versus minimum blasthole depth for quarries. (1980)

PROBLEMS

1. To determine the economics of a blast, we must evaluate drilling equipment which can produce holes with diameters from 1 to 4 inches. The rock has a specific gravity of 2.8 and a free flow or bulk-loaded ANFO with a specific gravity of 0.9 is to be used. Calculate the approximate burden in feet for the 1-, 3-, and 4-inch diameter holes.

2. The bench in the above problem is 16 feet high. For the 1-, 3-, and 4-inch diameter blasthole, determine the approximate amounts of
   a. stemming you would use
   b. the total depth of a blasthole (H)
   c. ANFO that would be required to load one hole.

3. Assume the four-inch blasthole is the most economic choice for Prob. 1 as far as drilling and explosive costs are concerned. What would you expect, based on considerations of stiffness, as to the
   a. rock breakage
   b. collar overbreak (backbreak)
   c. airblast
   d. leaving a toe.

4. Six-inch blastholes which are normally dry and loaded with bulk ANFO have been found to contain water. Therefore, they will be loaded with five-inch diameter water gel cartridges. The burden for the six-inch holes drilled is shale ($SG_r = 2.4$) is 13 feet.
   a. Determine the specific gravity of the slurry which would be necessary to result in breakage equivalent to dry holes loaded with ANFO.
   b. Determine the relative bulk strength of the product.

5. What would you expect as a charge in the 13-foot burden in Prob. 4 if severe geologic conditions, such as bedding steeply dipping into the cut and heavily cracked weakly cemented layers were encountered.

6. Determine the mean size of fragmentation for the following conditions:

Charge diameter (inches (decimal))	6
Burden (feet)	15
Spacing (feet)	19
Stemming	10
Subdrill (feet)	3
Bench height (feet)	60
Hole depth (feet)	63
Pattern factor (1— for square, 2— for staggered)	1
Number of rows	5
Explosive volume strength (ANFO = 100)	100
Specific gravity of explosive	0.80
Rock strength (weak-fissured = 1, strong-massive = 1.3)	4
Explosive charge per hole (lb)	490

7. Determine the fragmentation distribution for the conditions given in Prob. 6.

REFERENCES

ALLSMAN, P.L., "Analysis of Explosives Action in Breaking Rock," *Transactions of AIME,* 217:468–478, 1960.

ASH, R.L., "The Design of Blasting Rounds," In *Surface Mining,* E.P. Pfleider, (ed.), pp. 373–397, New York: American Institute of Mining Engr., 1968.

ASH, R.L., "The Influence of Geologic Discontinuities on Rock Blasting," Ph.D., dissertation, University of Minnesota, 1973.

ASH, R.L., "The Mechanics of Rock Breakage," Parts I, II, III, and IV. *Pit and Quarry,* 56, no. 2 (Aug. 1963), pp. 98–112; no. 3, (Sept. 1963), pp. 118–123; no. 4, (Oct. 1963), pp. 126–131; no. 5, (Nov. 1963), pp. 109–111, 114–118.

"Blast Fragmentation Prediction (Breaker)," Computer Software Package, Precision Blasting Systems, Montville, OH, 1987.

CUNNINGHAM, C., "The Kuz-Ram Model for Prediction of Fragmentation from Blasting," Preprint, First International Symposium on Rock Fragmentation by Blasting, Lulea, Sweden, 1983, pp. 439–453.

DICK, R.A., "Explosives and Borehole Loading," Subsection 11.7, *SME Mining Engineering Handbook,* A.B. Cummins and I.A. Given, (eds.), Society of Mining Engineers of the American Institute of Mining, Metallurgical, and Petroleum Engineers, Inc., New York, v. 1, 1973, pp. 11–99.

DICK, R.A., FLETCHER, L.R., AND D'ANDREA, D. V., "A Study of Fragmentation from Bench Blasting in Limestone at a Reduced Scale," U.S. Bureau of Mines, R.I. 7704, 1973.

HEMPHILL, G.B., "Blasting Operations," McGraw-Hill, New York, 1981, p. 258.

KONYA, C.J., "Current Blasting Practice Seminar," Precision Blasting Services, Morgantown, WV, 1972.

KONYA, C.J., "Proper Blasting Planning and Techniques," *Constructor Magazine*, March, 1976.

KONYA, C.J. and DAVIS, J., "The Effects of Stemming Consist on Retention in Blastholes," in, *Proceedings of the 4th. Conference on Explosives and Blasting Technique*, Society of Explosives Engineers, Morgantown, WV, pp. 102–112.

KONYA, C.J., and SKIDMORE, D.R., "Blasthole Depth and Stemming Height Measuring Systems," Final Report USBM Contract J0208022, 1981.

KONYA, C.J., WALTER, E.J., "Rock Blasting Manual," FHWA, Contract DTFH 61-83-C-00110, 1983, pp. 95–98.

KONYA, C.J., SKIDMORE, D.R., and OTUONYE, F.O., "Control of Airblast and Excessive Ground Vibration From Blasting by Use of Efficient Stemming," Washington, DC: U.S. Department of the Interior, Office of Surface Mining, 1981.

KUZNETSOV, V.M., Soviet Mining Science Vol. 9, no. 2, (1973), pp. 144–148.

OTUONYE, F.O., KONYA, C.J., and SKIDMORE, D.R., "Effects of Stemming Size Distribution on Explosive Charge Confinement: A Laboratory Study," *Transactions of the Society of Mining Engineers of AIME*, 1983.

PHUPHAIBUL, S., "A Fragmentation Study with Explosive Column Charges," MS thesis, West Virginia University, Morgantown, 1975.

PORTER, D.D., "Use of Fragmentation to Evaluate Explosives for Blasting," *Min. Cong. J.*, 60, no. 1, (Jan. 1974), pp. 41–43.

SPEATH, G.L., "Formula for Proper Blasthole Spacing," *Engineering News Record*, 218(3):53, 1960.

8

Blasthole Timing

INTRODUCTION

During the period from 1627 to early in the twentieth century, methods of accurate borehole timing did not exist. Long period delays were used with both cap and fuse and later with electric caps to sequence the holes in a designated manner. Sequencing holes proved to produce better results in fragmentation than shooting all of them at the same time.

The first millisecond delay, electric blasting caps, were produced in the late 1940s and quickly proved that vibrations could be reduced and fragmentation could be enhanced by using short period delays, rather that the long period delays previously available.

By the mid 1950s, most major explosives suppliers had millisecond delay blasting caps available that would sequence holes in fixed periods of milliseconds. The millisecond delay caps, both electric and nonelectric, were the most accurate method of timing until the mid 1980s.

It was readily apparent, however, that although we were timing blastholes in increments of thousandths of a second, results were variable. Fragmentation could change within a shot. Vibration levels from one shot to another could increase by two or three hundred percent.

The operator thought that the reason for this variability was strictly a function of the ground conditions, changing geology, wet holes or changes in types of explosives. Although these conditions could and were responsible for some of the variability, a great deal of the variability was caused by the inaccuracy of the initiators.

Operators, as well as regulatory agencies, simplified blasting calculations by assuming that the initiators would fire at the rated nominal firing times. Regulations and blast designs were, therefore, based on nominal firing times. What was not generally apparent was that the initiators often fired at times that were greatly different than their rated nominals. Caps of the same delay period would not fire at the same time. As an example, a blasting cap rated at 500 ms may fire between 530 to 570 ms. Because of this variability in firing time, blasthole timing from hole to hole would overlap and out of sequence firing would often result. Time overlaps caused out of sequence firing changing the confinement conditions on a borehole at the time it fired. This change in confinement or relief could cause significant changes in breakage, flyrock, airblast, and ground vibration.

Further research under highly controlled conditions, proved that accurate timing could enhance fragmentation and reduced the environmental effects of blasting. Technology advanced to the point where construction of more accurate initiators was feasible and the explosives industry increased the accuracy of their initiators.

Variability in firing times will always exist with pyrotechnic delays, however, new technology has reduced some of this variability, resulting in what is called "high-precision blasting caps." The use of high-precision caps, in some applications, will be a significant benefit, but only if the proper times are chosen for delays. High precision, in itself, is only a benefit if the specific cap times are matched to the conditions of that particular blast. The new generation of electronic detonators will give the user further control on blasting results, if properly used.

In order to better understand the importance of accuracy, let us look at Ex. 8.1.

Example 8.1

The pattern will consist of four holes in a single row. The initiation method will contain electric blasting caps and a sequential timer. The caps within the row are all of the same period of 500 ms, and a sequential timer setting of 10 ms between holes will offset the firing times of the caps. Figure 8.1 shows the pattern, as well as the sequential timer hookup.

If only nominal firing times are considered, the blast should function as shown in Fig. 8.1. The holes would theoretically fire independently with a 10-ms time delay between each and every blasthole.

If true cap data with actual nominal firing times and cap scatter are introduced for one manufacturer's millisecond delay caps, blastholes will not function as designed. Instead, the highest probability of overlapping of firing times of any two holes would be at 28% (Fig. 8.2). If high-precision cap data is used, the probability of overlap is reduced to 21%. If we change the caps within the blasthole from a delay of 500 to 1000 ms, the probability of overlap increases to 40% with regular ms caps and 28% with high precision caps. If we change, however, the 500 ms caps in the hole to 100 ms the probability of overlap for both caps is reduced (Table 8.1). This high probability of overlap shows what actually can occur, rather than what was intended by the designer. High probability of holes firing in a manner contrary to the design leads to problems of variability witnessed in actual blast. If the same pattern is fired using high-precision blasting caps, the probability of overlapping

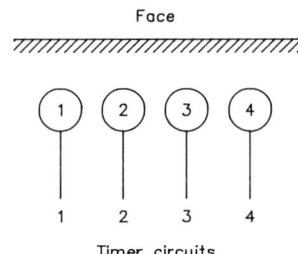

Timer circuits

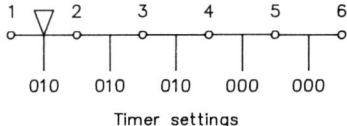

Timer settings

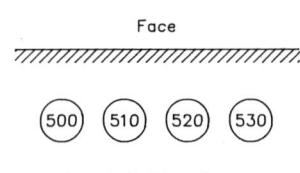

Nominal firing times

Figure 8.1   Pattern for Ex. 8.1.

Probability of overlap

Figure 8.2   Nominal firing times and probability of overlap.

TABLE 8.1   Typical Overlap Probabilities

Nominal Firing Time (milliseconds)	Millisecond Delay Probability of Overlap (percent)	High-Precision Probability of Overlap (percent)
100	11	1
500	28	21
1000	40	28

Note: Sequential timer setting is 10 ms between like period caps.

firing times of blastholes is reduced. Table 8.1 shows that the probability of overlap for Ex. 8.1 with different delay period in the hole. It shows that the probability of overlap is significantly reduced with low-period caps.

Initiation timing controls many of the results from blasting. It is as important to pick the proper timing between holes in a row and between rows, as it is to have the proper burden and spacing relationships. Accurate time selection is a new science and fundamental to producing reliable and predictable results from the blast.

CAP SCATTER

In any manufacturing process, the parts produced must be within given tolerances or they are rejected. Within the tolerance limits, parts will not be identically the same. In the manufacture of blasting caps, tolerances are held on the manufacturing process. Slight differences in caps result from the manufacturing process and cause times to vary slightly from one cap to another or from one batch to another. If we would test a statistically significant number of caps, we could determine the mean firing time for that particular batch and we could also determine the standard deviations or tolerances from the mean. In the blasting industry, these tolerances in firing time are called *cap scatter*. Blasting caps are given a rated firing time, for example 200 milliseconds, but this does not mean that from batch to batch the mean value of the firing time is 200 ms. It might be, for example, 195 plus or minus 5 ms. What this means is that it would be highly improbable for caps to fire at exactly the same time. Cap scatter, in itself is a problem, however, the manner in which the caps are used complicates the problem because cap scatter is overlooked in the design of the average blast. Operators design with nominal firing times and they assume that each and every hole fires at the nominal time. The problem is further complicated by the use of short delay times between charges. This is done by improperly using a sequential timer for electric shooting or with short period surface delays for nonelectric initiation. Short period surface delays often provide less time than the tolerance in firing time of the blasting caps inserted in the blastholes; therefore, what commonly occurs is crowding of firing times or out-of-sequence shooting. Out-of-sequence shooting causes problems when it occurs along a row even though relief is available at the burden. Significant problems of high vibration, blow out, and airblast occur when the out-of-sequence shooting occurs and back rows fire before front rows.

To properly design a blast, you must consider the affects of cap scatter and design within the tolerances of the cap. Experience has shown that too short a delay time between holes causes problems where as, a slightly longer time, normally does not. If you design on the worst case situation, where a certain minimum time window will be present either between holes or between rows, many of the common operational problems will disappear.

The US Bureau of Mines (1985) has done studies on cap scatter and some typical results are shown in Fig. 8.3. You will notice that when all caps are

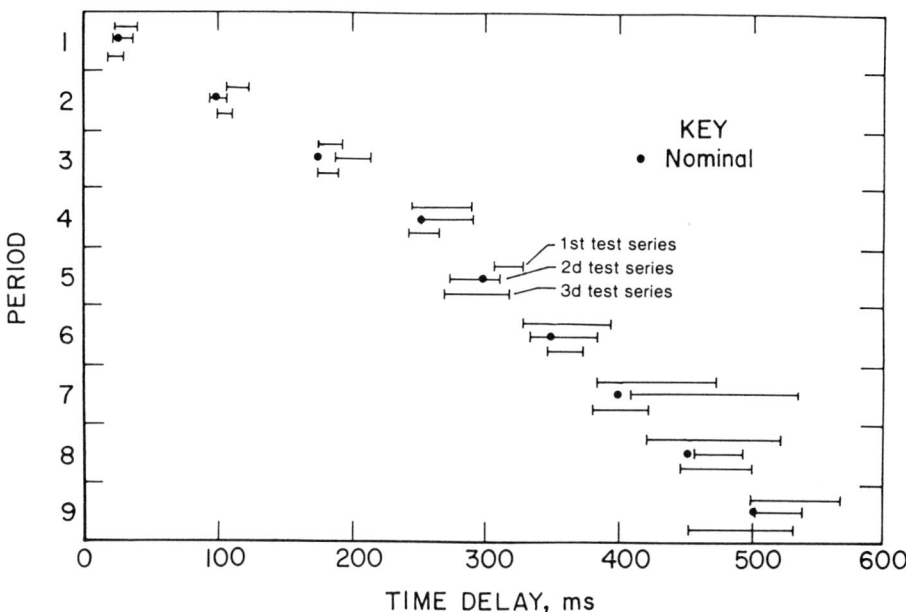

Figure 8.3 Cap tolerances for one manufacturer's caps.

energized simultaneously, some cap periods can overlap others, especially in the higher delay periods. In general, the higher cap periods produce a higher probability of overlap.

It is not uncommon to see blast designs where there are long period caps in the hole, with short surface delays. The rational for this is to get the entire system activated on the surface, in other words, get all caps burning in the hole before the holes start to fire. It is assumed that if this is not done, we stand a chance of getting breakage in the surface lines or cutoffs of initiator leads before they energize. It is the authors' experience that cutoffs are blamed for many other problems such as poor explosive, poor design, or poor initiation of the blast. If you consider the time delays before rock motion starts after detonation and the trajectories at which broken rock moves, cutoffs from flyrock are virtually impossible on the average blast. Uplift of the rock surface as a result of out of sequence firing can cause cutoffs in the blasthole. The use of fully activated systems has caused a number of problems with out-of-sequence shooting because the long delay period used in the holes creates a great deal of cap scatter and a high probability of overlap. The surface time delays between the holes is insufficient to overcome the cap scatter problem.

A more logical solution to the scatter problem is to use short delays within the holes and a longer delay period on the surface. A common rule of thumb is that if the adjacent hole in any direction is energized before a hole fires, cutoffs should not be a problem.

Methods to overcome problems with cap scatter.   We must know
what amount of cap scatter exists for each cap period to fully control cap scatter. We
must also know the true mean firing time. If this information is available we can
calculate the probability of adjacent holes or adjacent charges overlapping in firing
times. If the probabilities are high we can change our design before we fire the shot.
Many commonly used designs have a probability of overlap of 45 to 48%, which
means that there is a high probability that the blast will never function as designed
on paper.

Another manner, in which we can handle the scatter problem, if the true mean
firing time and scatter data are not available, is to first assume that the cap mean
firing time will be near the rated firing time and use a cap scatter of 10% of the cap
firing time for standard ms caps. Using these assumptions, we can calculate the
probability of cap overlap of adjacent charges.

TIMING EFFECTS ON FRAGMENTATION

Two general conditions of initiation timing will be discussed. Blastholes can be
fired near simultaneously or fired delayed from one another. Simultaneous initiation
along a row requires a larger blasthole spacing than firing on delays, and therefore,
since holes are spaced further apart, the cost per cubic yard of broken material is
reduced. One major drawback of having simultaneous initiation along a row, is
increased ground vibration since many holes fire at the same time. Although more
yardage is produced by using instantaneous initiation and increased borehole spacing, the fragments would be larger than that produced by proper delay initiation
timing. Delay timing reduces ground vibration and produces finer fragmentation.

Hole-to-hole delays.   There are some relatively simple procedures to
calculate the time delay for delay initiation hole-to-hole along a row. Table 8.2
supplies time constants for various rock types. The information in this table can be
used along with the following equation:

$$t_h = T_H \times S \qquad (8.1)$$

where

t_h = hole-to-hole delay in ms
T_H = delay constant hole-to-hole from Table 8.2
S = spacing in ft

When rock is highly fractured the delay constant given in Table 8.2 may be
increased by as much as 50% to accommodate local geologic conditions.

Row-to-row delays.   Guidelines for row-to-row initiation are as follows:

a. Short delay times cause higher rock piles close to the face (Fig. 8.4).
b. Short delay times cause more endbreak.

TABLE 8.2 Time Delay Between Blastholes

Rock Type	T_H Constant (ms/ft)
Sands, loams, marls, coals	1.8–2.1
Some limestones, rock salt, shales	1.5–1.8
Compact limestones and marbles, granites and bassalts, quartzite rocks, gneisses and gabbroe	1.2–1.5
Diabase, diabase porphyrites, compact gneisses and micashists, magnetites	0.9–1.2

    c. Short delay times cause more violence, airblast, and ground vibration.

    d. Short delay times can create flyrock.

    e. Long delay times decrease ground vibration.

    f. Long delay times decrease backbreak.

    To determine the delay time to be used between rows in production blasts, the general guidelines are given in Table 8.3.

    Delay times in general should not be less than 2 ms per ft of burden between rows. Delay times should normally be no greater than 6 ms per ft of burden between rows unless rock casting is desired. When wall control is critical in multirow shots (6 or more rows), row-to-row delays may be expanded to as much as 10 to 20 ms/ft

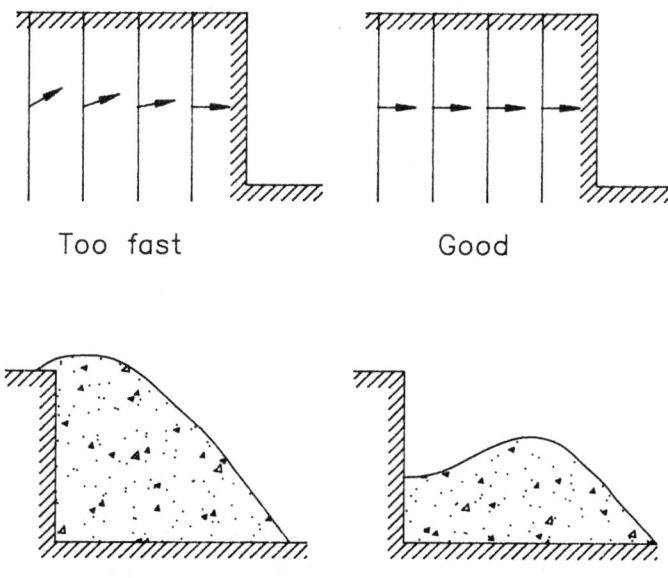

Figure 8.4 Rock displacement as a function of timing.

TABLE 8.3   Time Delay Between Rows

T_R Constant (ms/ft)	Result
2	Violent excessive airblast, back-break, etc.
2–3	High pile close to face, moderate airblast, backbreak
3–4	Average pile height, average airblast and backbreak
4–6	Scattered pile with minimum backbreak
7–14	Blast casting

of burden to obtain low muck piles. An equation to determine delay time between rows is as follows:

$$t_r = T_R \times B \qquad\qquad (8.2)$$

where

$$
\begin{aligned}
t_r &= \text{time delay between rows in ms} \\
T_R &= \text{time constant between rows} \\
B &= \text{the burden in ft}
\end{aligned}
$$

The selection of an approximate time delay in milliseconds for a test blast is accomplished by determining a time factor using the tables and making one multiplication. The values obtained, however, may be difficult if not impossible to implement in the field because of the limitations in hardware available from the manufacturers and because of ground vibration considerations.

Many problems which result from blasting and cause airblast, flyrock, excessive vibration, and poor fragmentation, are directly related to the initiation timing (Fig. 8.5). Tables 8.2 and 8.3 help us to determine timing values, which can be used to better our blasting performance. Timing must also be considered for its potential to cause environmental problems.

It is generally proposed by regulatory agencies that charges be fired with at least 8 ms delay between charges if they are to be considered independent events from the standpoint of ground vibration. This simple approach, however, is misleading and does not provide satisfactory results in environmental control or fragmentation. Blasthole firing times must be carefully selected and scaled to the other pattern dimensions.

SELECTION OF PROPER TIMING

In the previous sections, we discussed some of the physical effects of timing. To control blast effects, we must fire holes within a certain time period which we will call *timing windows*. Unfortunately, these windows are not the same for all effects

Figure 8.5   Holes venting from improper design. (Courtesy of Austin Powder Company)

and we must find a window, if possible, that will control all of the effects that we have discussed. Timing becomes even more complicated since the amount of time to control these different effects, is dependent on two major factors. The first is the rock type, but more importantly are the physical dimensions of the shot, the burden, and the spacing. It should be obvious, that if we want to retain the same effects from one blast to another and we double the burden, we must also double the timing. The reason for this is that cracks travel at a certain maximum speed in the rock mass, which cannot be changed by increasing or decreasing the amount of explosive.

The initiators provided by the manufacturers are produced with fixed time periods. Often operators use these same periods regardless of the size of the blasthole or pattern dimensions. The size of the breakage can be different, if in one case we have 3-in. holes with 6-foot burdens, while in another case we have 15-in. holes with 30-foot burdens and the timing is identical in both cases.

It is a common practice on large-hole strip mine blasts to increase time intervals over what would be used on similar blasts with smaller holes because better control and fragmentation results. Experience has shown that scaling time to physical dimensions does produce better results.

With the supplies available from the manufacturer, we many not be able to select the absolute best time for every situation, even if we knew what those best times were. We would at best be able to get "close."

Implementation problems.   One of the problems in implementation of proper timing, is that we often look at timing in a simplistic fashion. We make assumptions that detonators will fire at their rated firing times and we lay out patterns based on the rated firing times. We commonly see that first rows of patterns function properly, but as later rows fire, we see holes that do not function as anticipated, firing either too soon or too late. Fragmentation can be larger and final walls are not uniform. The signs are all there, but we often have not interpreted these signs properly. We commonly blame problems, that result in blasting, on ground conditions. Although ground conditions do effect the blast, normally they're the scape goat for problems which occur from poor drilling, which changes the timing effects between holes and problems which occur with the timing itself. We must design patterns realizing that blasting caps do not normally fire at exactly the rated firing time, they may either fire sooner or later. We must consider cap scatter in design. Cap scatter can change the actual number of milliseconds between holes and between rows in the shot. In general, the greater the delay period, the greater the milliseconds of cap scatter.

Timing calculations.   In order to approach the problems of timing in a systematic scientific manner, the expected outcomes or potential problem areas must be evaluated in a step-by-step fashion before the timing system can be determined. Questions must be asked and the answers prioritized based on the desired final outcome. Such questions are as follow:

1. Is muck pile height a consideration? If so, are we trying to pile the material, scatter the material, or deliberately cast the material?
2. Is wall control a factor? If so, is it more or less important than the piling and casting considerations?
3. Is the sizing of the rock important? Is uniformity important or only average size? Do we want to produce rip-rap in the blast?
4. Is airblast a concern? If so, are we concerned with blowout from back rows causing airblast or are we concerned with timing causing enhanced concussion, from the rock face falling.
5. Are we concerned about flyrock traveling considerable distance, especially from back rows in a blast.
6. Is maximum vibration level a problem, or are we far enough removed from residents that we don't have to worry about effects of vibration? If vibration is a problem, what standard are we using to measure vibration? Are we con-

TABLE 8.4 Timing Control Functions

1. Rock Placement Considerations
   a. high pile close to face
   b. average pile
   c. scattered pile
   d. rock casting
2. Wall Control Is Important
3. Fragmentation Sizing Desired
   a. average
   b. best possible
4. Airblast
   a. back holes venting important
   b. wall collapse concussion important
5. Flyrock Control Important
6. Blast Vibration Important
   a. 8-ms legal limit
   b. few nearby homes
   c. homes surrounding blast site

cerned with the 8 ms legal limit or are we truly concerned with maximum vibration levels, which would be measured by a seismograph? Are we concerned with the formation of *hot spots* in the area, some areas which will receive considerably higher vibration levels than the average?

These general questions will have to be answered before we can proceed with the best timing layout. You can address these potential problem areas by considering the list given in Table 8.4. The potential problem areas are called timing control functions. The timing control functions can be associated with time delay intervals scaled to pattern dimensions.

The timing control functions from Table 8.4 are identified for the specific site and blast. The functions are prioritized. Table 8.5 is a listing of time windows to coincide with the timing control functions of Table 8.4. Time lines can then be established and "best" time windows identified. Example 8.2 illustrates the step-by-step process in detail.

TABLE 8.5 Selection of Time Windows

Functions	Time Windows
1A	2–3 ms per ft of burden
1B	3–4 ms per ft of burden
1C	4–6 ms per ft of burden
1D	7–14 ms per ft of burden
2	3–14 ms per ft of burden
3A	0–5 ms per ft of spacing (or more)

(continued)

TABLE 8.5   (*Continued*)

Functions	Time Windows
3B	1–2 ms per ft of spacing massive (depending on rock type, 3–4 ms per ft of spacing for highly fissured)
4A	2 or more ms per ft of burden between rows
4B	Direction of initiation along row, adjacent holes 0.8 ms $> t > 1$ ms per ft of spacing
5	2 or more ms per ft of burden between rows less than 25 ms per ft of spacing
6A	8 ms nominal times between delays
6B	Tune blast to minimize vibration in one direction. Danger between 0.1–1.0 ms per ft of burden or spacing, whichever significant (include cap scatter considerations)
6C	Time (including scatter) greater than 1 ms per ft to hole on next delay

Note: Row-to-row time should be at least twice as great as hole-to-hole so that proper relief will be available for subsequent rows.

Example 8.2 Pattern Information

A massive limestone formation is blasted with 6 rows of holes with 10 holes per row. The drill pattern is accurately drilled with 4-in. holes on a 10- by 12-foot pattern, the shot cannot have more than 1 hole firing at a time since homes surround the operation. The pattern will open at only one location and it will be a row by row pattern.

Airblast, flyrock, and blowout must be kept to a minimum. Wall control is an important consideration, vibration levels are also of concern.

Functions	Time Windows
2	3–14 ms per ft of burden between rows
4A	2 or more ms per ft of burden between rows
5	2 or more ms per ft of burden between rows less than 25 ms per ft of spacing along row
6C	Time (includes scatter) greater than 1 ms per ft to next delay
3A	0–5 ms per ft of spacing along row
1B	3–4 ms per ft of burden

Solution

1. Select and prioritize function
2. Order of priority are 2, 4A, 5, 6C, 3A, 1A
3. Establish prioritize listing with time windows
4. Multiply time window by actual pattern dimension and fill in time lines (Fig. 8.6).
5. Evaluation of time diagrams

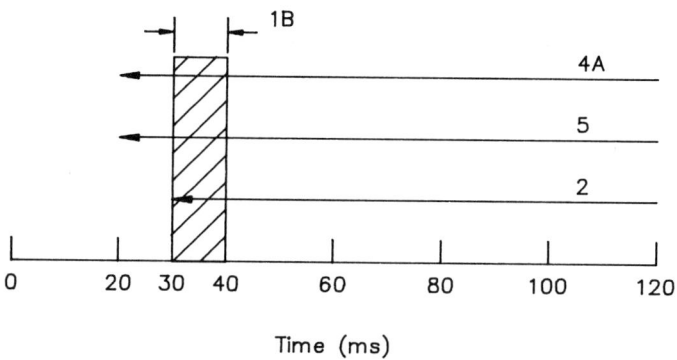

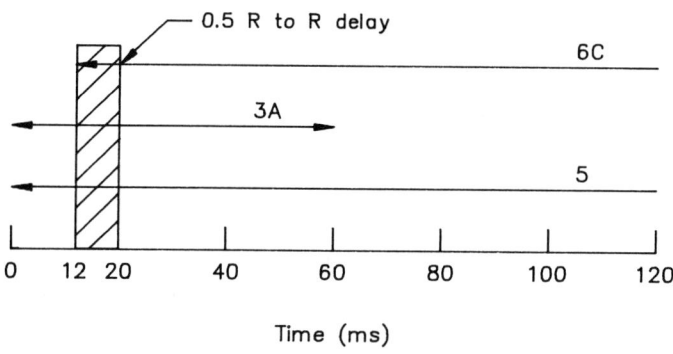

Figure 8.6 Timing window selection.

The row-to-row window is most critical giving a 30 to 40 ms row-to-row time as best to meet all criteria.

The time window along the row could be between 12 to 60 ms, however, if this is a row-by-row pattern and we want relief at time the second row shoots, time should be no greater than half the time selected row-to-row. Therefore, if for example, 40 ms was selected between rows then time along row would be minimum 12 ms and maximum 20 ms.

These times should be available after we consider the cap scatter of the specific initiators.

Blasthole Timing for Vibration Control

The vibration wave form generated from a blast is a composite of the single wave forms generated as each individual charge fires. If all charges were identical in size, geometry, and confinement, one could use linear superposition to define the final wave form. Linear superposition is a method that has been used by the geophysical industry for many years in oil and gas exploration to develop synthetic seismograms. One reason linear superposition models often do not work for blasting applications is because not only are the charge geometries (confinement) different but also because timing varies from blasthole to blasthole. Blasthole timing accuracy is important, however, it is equally important to select the proper milliseconds of delay between holes to produce the desired end results.

It is common practice in field blasting to monitor the vibration directly behind the shot. The assumption is made that it will be the direction of maximum vibration. The direction of maximum vibration cannot be assumed to be behind the shot since the direction is a function of the blasting pattern, rock type, and the initiation timing. It is not uncommon to find the direction of maximum vibration off the end of a shot in the direction in which the initiation is traveling through the pattern. This vibration level can be many times that recorded behind the blast, changing nothing but the initiation timing in milliseconds between holes can result in significant reduction in the vibration level in any direction. Minimizing the vibration in one direction, however, may increase it in another. It is therefore, necessary to consider vibration levels in all directions before you can be sure that the optimum results have been obtained.

The use of timing to control maximum vibration can be effective, however, the use of timing to significantly change frequency content of the wave is much more difficult. When monitoring at structures close to blasting sites, the waves tend to be of higher frequency than when monitoring at great distances from the blast as is commonly done in coal strip mine blasting. When monitoring at distances measured in thousands of feet from the shot, most of the high-frequency components vanish and only low frequency remains. This is true in spite of use of superposition models which would predict a great deal of high-frequency in the seismic wave. The reason the high-frequency components vanish is because high frequency is readily attenuated by the rock mass itself and only low frequency remains at great distances from the blast. At distance, it is the rock mass that controls frequency not specifically the delay times.

When blasting is close to structures such as in construction blasting, you can design a shot where high frequencies dominate. However, at far distances we should be looking at designing blasts to minimize peak particle velocity since we have little control on frequency.

Simple linear superposition models. Simple superposition models use the vertical, longitudinal, and transverse wave forms generated by a single charge firing. The wave form for each individual component is superimposed on itself at different delayed times. The single hole wave trace for each component is obtained by seismic monitoring at the structure needing protection.

Directional vibration effects. In order to better understand the effects of timing on vibration, we will examine data from a quarry blast. To demonstrate the principles, the pattern will only consist of 4 holes with 500 lb of explosive per hole. The blastholes will be drilled with a 15 ft burden and 20 ft apart. We will use a commonly used timing sequence with electric blasting caps and a sequential timer. The blastholes will each contain a 500-ms electric blasting cap. The time interval set into the sequential timer will be 9 ms. The nominal firing times of the blasthole will be 500, 509, 518, and 527 ms (Fig. 8.7).

A single hole test shot will be used to generate a signature wave form. We will examine only a single component trace for this example. The single hole was monitored at a location 1000 ft from the blasthole. The wave form for the single hole is the bottom trace on Fig. 8.8. The center trace on Fig. 8.8 indicates the time in milliseconds. If we use a simple linear superposition, we can produce a wave form as shown on the top trace of Fig. 8.8. The signature wave form from a single hole would have a peak particle velocity of 0.06 in. per sec. If we use a superposition model to project the maximum value 1000 ft away along the line of the holes in the direction towards which initiation is proceeding, we would anticipate a peak particle velocity of 0.098 in. per sec. The monitoring location for this example would be shown as location A on Fig. 8.9. The value at location A is not significantly more for 4 holes than it would have been for 1 single hole.

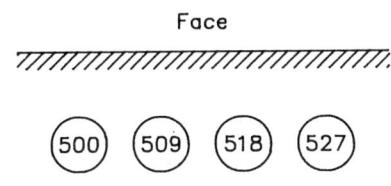

Figure 8.7   Blasthole nominal firing times.

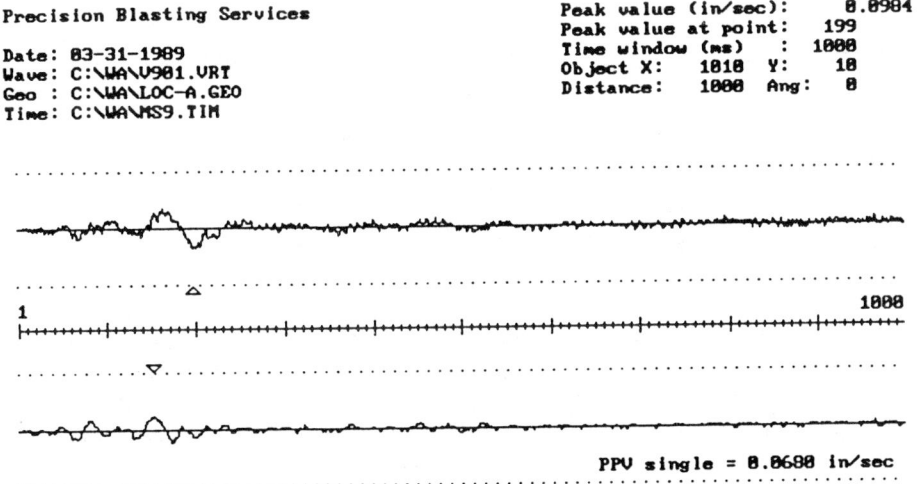

Figure 8.8   Simple linear superposition.

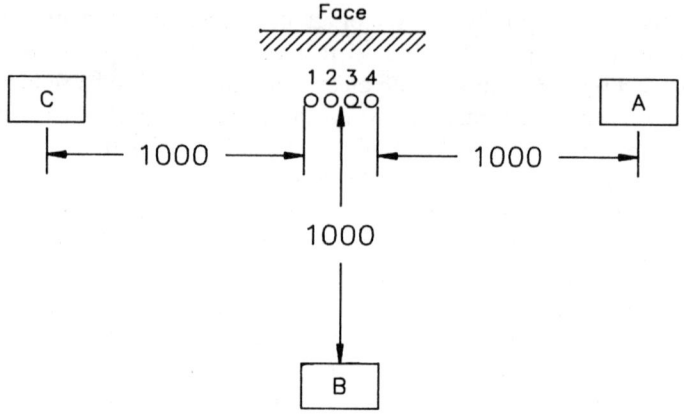

Figure 8.9   Monitoring locations at quarry.

Let us now use a more complex superposition model called Pre-Seis (1988), to compare the results with the simple superposition model. The Pre-Seis model contains information not only on timing but also on the blasthole and pattern geometry. The information needed for this model is as follows:

LISTING OF INFORMATION NEEDED

1. Burden	**14.** Number of primers
2. Spacing	**15.** Primer composition
3. Subdrilling	**16.** Boosters
4. Stemming depth	**17.** Geologic factors
5. Type of stemming	**18.** Number of holes in a row
6. Bench height	**19.** Number of rows
7. Number of decks	**20.** Type of initiator
8. Charged geometry	**21.** Row-to-row delays
9. Powder column length	**22.** Inhole delays
10. Rock type	**23.** Initiation accuracy
11. Rock physical properties	**24.** Distance to structure
12. Explosive energy	**25.** Face angle to structure
13. Actual delivered energy	**26.** Signature wave

With the use of this additional information, the results obtained are quite different than with the simple superposition model. Instead of the peak value being 0.098, the peak value is 0.192 at point A on Fig. 8.9, showing that the conditions are not simple as assumed in the simple linear superposition model. Figure 8.10 shows the resulting wave form on the top trace.

Let us use the Pre-Seis model to predict the wave form and peak particle velocity 1000 ft directly behind the blast at location B to illustrate the effects of

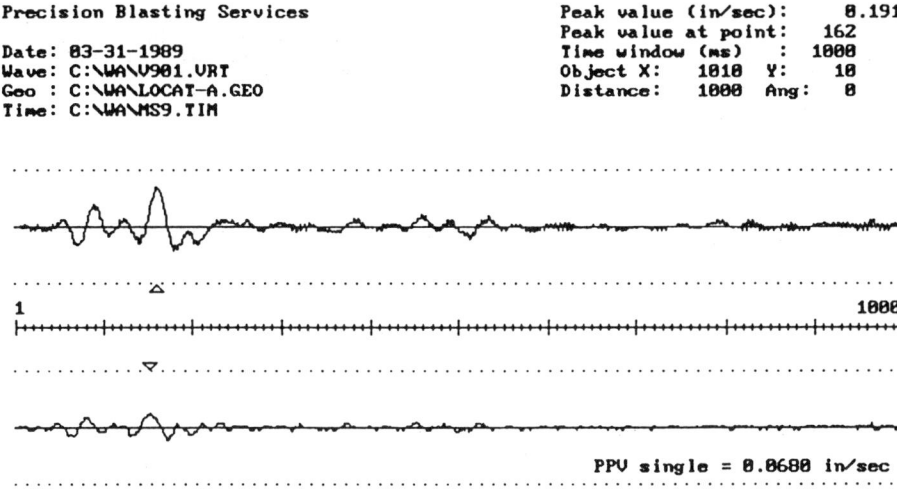

Precision Blasting Services
Date: 03-31-1989
Wave: C:\WA\V901.VRT
Geo : C:\WA\LOCAT-A.GEO
Time: C:\WA\MS9.TIM
Peak value (in/sec): 0.1919
Peak value at point: 162
Time window (ms) : 1000
Object X: 1010 Y: 10
Distance: 1000 Ang: 0
1
1000
PPV single = 0.0680 in/sec

Figure 8.10 Pre-Seis simulation for location A, 1000 ft from the east side of the blast.

pattern geometry. We will use the same signature wave. Figure 8.11 shows that the peak particle velocity behind the blast would only be 0.098 or half of what was obtained on the end.

If we change our monitoring location from 1000 ft in the direction of detonation at point A to 1000 ft in the opposite direction from detonation, point C on the west end, we would get the results as shown in Fig. 8.12. At location C, we would have 0.078 or nearly the same value of peak particle velocity as we would have for a single hole blast.

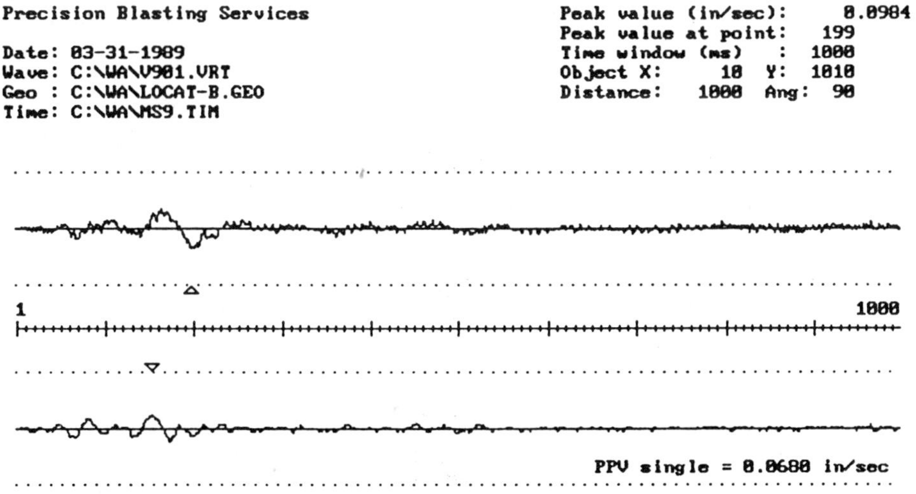

Precision Blasting Services
Date: 03-31-1989
Wave: C:\WA\V901.VRT
Geo : C:\WA\LOCAT-B.GEO
Time: C:\WA\MS9.TIM
Peak value (in/sec): 0.0984
Peak value at point: 199
Time window (ms) : 1000
Object X: 10 Y: 1010
Distance: 1000 Ang: 90
1
1000
PPV single = 0.0680 in/sec

Figure 8.11 Pre-Seis simulation for location B, 1000 ft behind the blast.

Precision Blasting Services

Date: 03-31-1989
Wave: C:\WA\V901.VRT
Geo : C:\WA\LOCAT-C.GEO
Time: C:\WA\MS9.TIM

Peak value (in/sec):   0.0784
Peak value at point: 216
Time window (ms)   :   1000
Object X: -990   Y: 10
Distance: 1000   Ang: 0

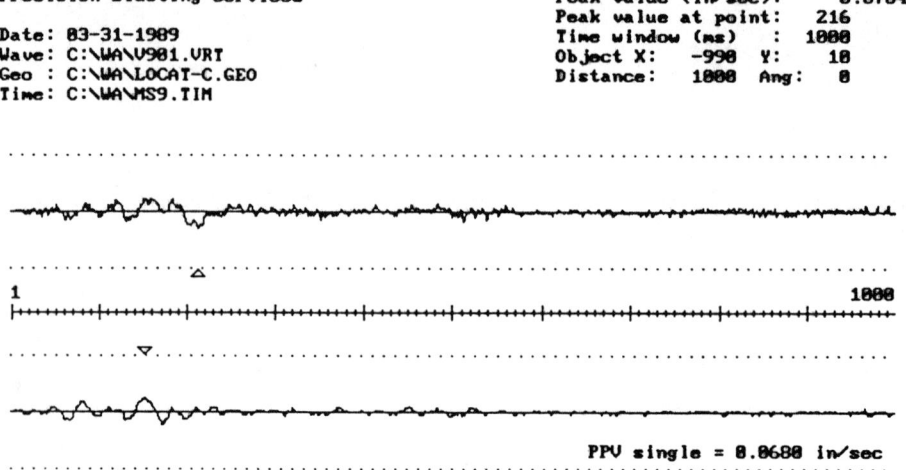

PPV single = 0.0600 in/sec

Figure 8.12 Pre-Seis simulation for location C, 1000 ft from the west side of the blast.

If we summarize this data on Fig. 8.13, we can see the directional vibration effects which result from the timing sequence selected. The maximum vibration at the same distance from the blast is at point A where the vibration level is 2.5 times the vibration level at point C. It is obvious that we can no longer make the assumption that at equal distances from the blast, we can anticipate equal vibration levels.

To further demonstrate the importance of the proper time selection, let us take the same example and change the time delay between holes from 9 ms to 17 ms. The 4 blastholes would have relative firing times of 0, 17, 34, and 51 ms. If we could compare anticipated vibration levels at point A which was the maximum using the shorter delay periods, we find a significant reduction in vibration as shown in

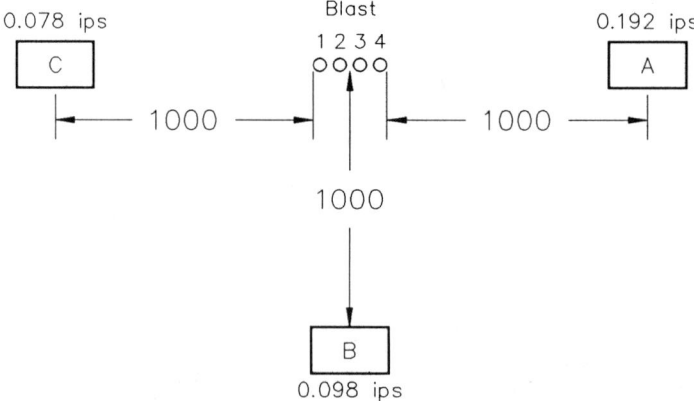

Figure 8.13 Vibration levels surrounding blast site.

Fig. 8.14. At point A with the 9-ms delays between holes, we had a 0.192 in. per sec maximum, while with the 17-ms timer interval, we had a maximum of 0.078 in. per sec.

Blast design parameters in superposition models.   To properly conduct a vibration simulation, we must include the blast design parameters. We commonly have assumed that a given number of pounds of explosives can be related to a specific vibration level. That assumption is an oversimplification of what truly occurs. For example, let us assume we are going to shoot 1000 lb of explosive. That 1000 lb of explosive could be in a very large diameter hole of very short length. For example, if that 1000 lb was bulk loaded into a 15-in diameter blasthole and the explosive had a density of 1.3, we would be loading 97.5 lb per linear ft of blasthole or we would have a total charge length of 10.25 ft. If that same 1000 lb of explosive was loaded into a 6-in. blasthole, we would have a loading density of 15.6 lb/ft or a total charge length of 64 ft. The 10.4 ft length of charge will detonate much quicker than the 64 ft length. Therefore, it would generate more vibration.

The degree of confinement on a charge also influences the amount of vibration produced. This is why blastholes that have excessively large burdens produce much more vibration than similar blastholes fired with very small burdens. Blastholes that prematurely relieve themselves by blowing out as a result of insufficient stemming can also produce more vibration than those that are confined and break the rock properly. All of the blast parameters influence the confinement of the hole at the time it shoots. The degree of confinement controls the amount of energy that goes into useful work energy or into waste energy.

If all factors could remain constant, a single simulation may predict the anticipated wave trace and peak particle velocity. For example, Fig. 8.15 shows the

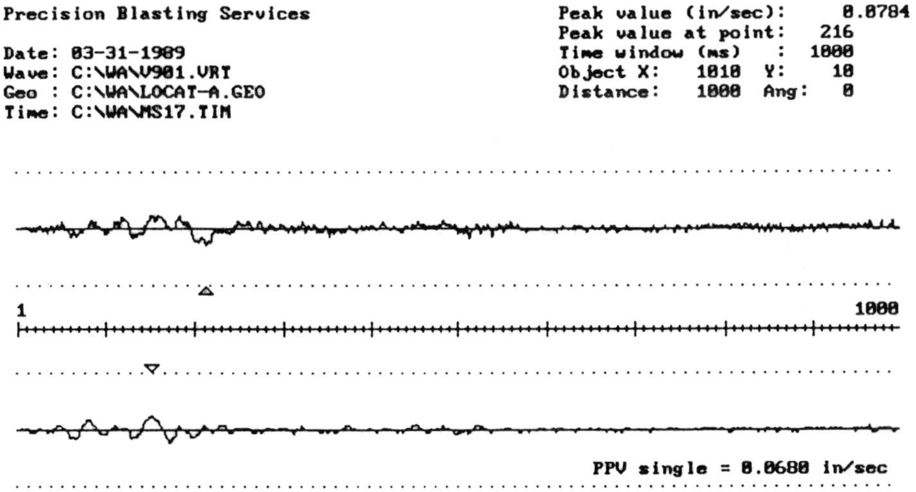

```
Precision Blasting Services              Peak value (in/sec):      0.0784
                                         Peak value at point:      216
Date: 03-31-1989                         Time window (ms)   :      1000
Wave: C:\WA\V901.VRT                     Object X:    1010  Y:      10
Geo : C:\WA\LOCAT-A.GEO                  Distance:    1000  Ang:     0
Time: C:\WA\MS17.TIM
```

```
                                                    PPV single = 0.0680 in/sec
```

Figure 8.14   Pre-Seis simulation with 17 ms delays at location A.

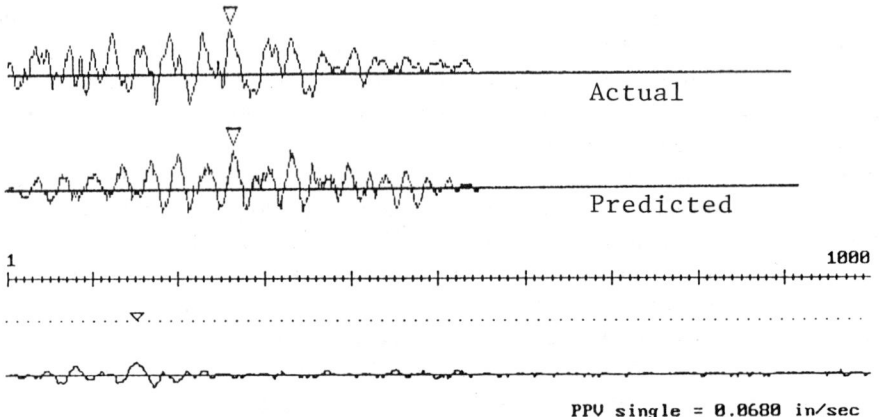

Precision Blasting Services   Peak value (in/sec):   0.2183
                                  Peak value at point:
Date: 03-31-1989   Time window (ms)   :   1000
Wave: C:\WA\V901.VRT   Object X: 1300 Y: 10
Geo : C:\WA\G10-1-3.GEO   Distance: 1290 Ang: 0
Time: C:\WA\T10-1-3.TIM

Actual

Predicted

1   1000

PPV single = 0.0680 in/sec

Figure 8.15   Predicted and actual wave traces using Pre-Seis simulation.

predicted wave trace generated by using the Pre-Seis analysis technique and the actual wave trace which resulted from the subsequent blast.

On the average blast, it is impossible to hold all factors constant. Not even the ground characteristics remain constant. We can place tolerance levels on all the blasting parameters and rock properties. We can include these tolerances in our calculations to more closely project the anticipated range of vibration levels. If the blast design parameters are realistically assessed with the true tolerances, each wave trace would be different than the previous. A statistical analysis can be conducted on the results of many simulations and a normal distribution of the PPV can be projected as shown in Fig. 8.16. Using this data an operator can determine the most likely level of vibration and the plus or minus tolerances which we could anticipate from one blast to another.

The Importance of Timing for Overall Blast Design and Vibration Control

Blast results depend on the selection of proper time periods and the ability to achieve those time periods in the field. All initiators have built in tolerances on time. These tolerances must be considered in the design. There are many ways to achieve what seemingly is the same end results and yet some methods would have a high probability of failure. Let us assume it is our goal to have holes sequencing at 17-ms intervals. Let us look at a simple pattern as shown in Fig. 8.17 which consists of 5 holes in a single row. We will use Nonel as the initiator with 17-ms surface

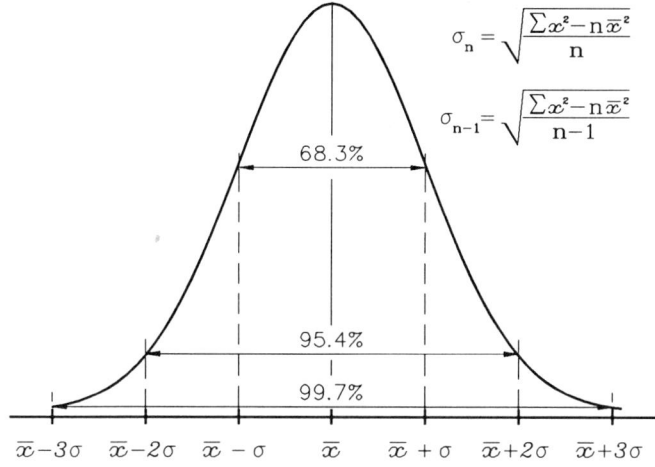

$$\sigma_n = \sqrt{\frac{\sum x^2 - n\,\overline{x}^2}{n}}$$

$$\sigma_{n-1} = \sqrt{\frac{\sum x^2 - n\,\overline{x}^2}{n-1}}$$

68.3%

95.4%

99.7%

$\overline{x}-3\sigma \quad \overline{x}-2\sigma \quad \overline{x}-\sigma \quad \overline{x} \quad \overline{x}+\sigma \quad \overline{x}+2\sigma \quad \overline{x}+3\sigma$

Figure 8.16   Statistical analysis of data using mean values and tolerances.

delays along the row. We will use the same cap period in the bottom of every hole, therefore theoretically the sequencing should be controlled by the surface delay alone. We will compare two timing sequences. The first will contain a 100-ms downhole delay and the second will contain a 200-ms downhole delay (Fig. 8.18). Both hookup methods with theoretically sequenced holes 17 ms apart. For the purposes of this example, we will assume that one standard deviation is 5% of the cap period. That is to say that the 100-ms cap could fire at 100 plus or minus 5 ms. The 200-ms primadet could fire at 200 plus or minus 10 ms.

Commercial software packages are available to calculate firing times of holes for different patterns and the probability of overlap. We will use one of these software packages called "Quartz" (1987) which analyzes time delays for calculations in this example. Figure 8.19 is a summary of one of the output pages from this software. The 100-ms period primadet produced a 0.72% probability of overlap between charges. The 200-ms downhole delay produced a 11.44% probability of overlap or approximately 16 times higher probability of failure than the previous case. What this means in practical terms, is that statistically the 200-ms downhole delay could overlap approximately 1 out of 9 times. There are many combinations of electric or nonelectric delays that would theoretically sequence holes at 17 ms, however, the probability that holes would overlap could even be larger with other combinations.

Before selecting any combination of initiators which theoretically would produce the desired time interval, we must determine whether practically those time

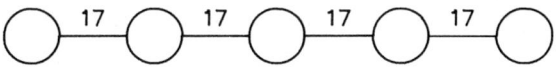

Figure 8.17   Single row Nonel hook-up for trunkline delays.

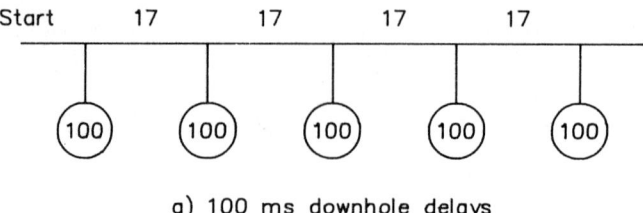

a) 100 ms downhole delays

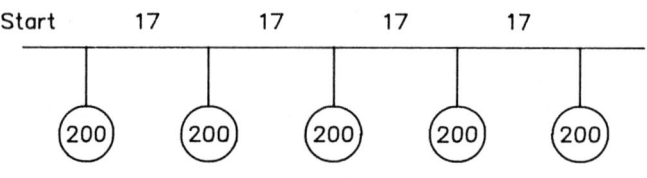

b) 200 ms downhole delays

Figure 8.18 Hook-up for 100 and 200 ms downhole delays.

Hole #	Probability (100 ms)	Probability (200 ms)
1		
	0.72 %	11.44 %
2		
	0.72 %	11.44 %
3		
	0.72 %	11.44 %
4		
	0.99 %	11.44 %
5		

Figure 8.19 Probability of overlap for 100 and 200 ms downhole delays.

intervals will be achieved. We can do this by doing hand calculations of overlap probability or by using one of the software packages available to us.

PROBLEMS

1. Define the term cap scatter.
2. Why is cap scatter important?
3. Determine the best time delay along a row of holes with a spacing distance of 5 ft in shale. Determine the proper timing for hole-to-hole delays on a blast which has spacing distances of 40 ft in shale.
4. You had been asked to design a blast which contains 6 rows of holes. Determine the

minimum time delay between rows if the distance between rows is 10 ft. Determine the minimum time delay between rows if the distance between rows is 30 ft.

5. We are to design the timing for a blast which has a 10 ft burden and 15 ft spacing. We are concerned with maintaining wall control. Determine the best time window between rows for a multirow blast.

6. Assume you are planning to use 800-ms caps in the blastholes with a 10-ms sequential timer setting between holes in a row. What is the probability of overlap of any 2 holes? What is the probability of overlap if the time delay is set at 25 ms between holes?

7. Example 8.2 indicates that time windows between rows should be between 30 and 40 ms and the time between holes in a row would be between 12 and 20 ms. Assume that you would use 17-ms surface delays with Nonel between holes in a row and 42 ms between rows with a 200-ms delay in the hole. The blast has 4 rows with 5 holes per row. Determine the firing times of all holes. Determine which holes have the maximum probability of causing an overlap and calculate the actual probabilities.

8. The Ajax Construction Company is blasting in shale. They are using a 2-in. diameter hole in a pattern with a 6-ft burden and 8-ft spacing. The blast contains 3 rows of holes. Each hole is fired on an independent delay.
   a. What would be the delay time hole-to-hole within a row?
   b. What would be the delay time row-to-row?

9. A strip mine has a massive rock overburden drilled with 6 rows of holes, with 8 holes per row. The drill pattern is accurately drilled with 12-in blastholes on a 32-ft burden and 32-ft spacing. The shot can not have more than 1 hole firing at a time, since there are homes very close to the blasting operation. The pattern can only open at one location and will be a row-by-row pattern. Assume that there is a 10% cap scatter. The pattern is meant to break the rock and restrain as much motion as possible. On the other hand, flyrock and blowout can not be tolerated. Airblast and ground vibration levels are of concern as well as fragmentation.
   Determine the time windows hole-to-hole and row-to-row which will best meet the above criteria.

10. You will be blasting at a quarry that has a 185-ft high bench. You will be shooting a single row shot along the wall and are concerned about concussion from the falling wall. The blastholes are drilled on a 20 by 27-ft pattern. What is the minimum delay time between holes, such that concussion from the falling wall would not become additive hole-to-hole.

REFERENCES

BAJPAYEE, T.S., MAINIERO, R.J., and HAY, J.E., "Overlap Probability for Short-Period-Delay Detonators used in Underground Coal Mining," US Bureau of Mines, RI 8888, 1985.

KONYA, C.J., "Addendum—Rock Blasting Manual," FHWA Contract DTFH-61-83-C-00110, Washington, 1986.

"Pre-Seis Computer Analysis," Precision Blasting Services, Montville, OH, 1988.

"Time Delay Analyzer (Quartz Series)," Computer Software, Precision Blasting Systems, Montville, Ohio, 1987.

WINZER, S.R., "The Firing Times of MS Delay Blasting Caps and Their Effect on Blasting Performance," Prepared for National Science Foundation (NSF Apr. 77-05171), Martin Marietta Laboratories (Baltimore, MD), June 1978, pp. 36; available for consultation at Bureau of Mines Twin Cities Research Center, Minneapolis, MN.

9

Blasting Pattern Design

PRINCIPLES OF PATTERN DESIGN

A blasting pattern is constructed from properly designed single blastholes which are placed into a geometrical relationship with one another and with the open face. The selection of the spacing distance between blastholes in a single row is dependent upon the initiation timing of the adjacent holes and the stiffness ratio (L/B).

If holes in a row are initiated near simultaneously, spacings must be spread further apart than if adjacent holes are fired delayed from one another. If holes are spaced too close together, a number of undesirable effects occur. Cracks from the closely spaced blastholes will link prematurely causing a shattered zone in the wall between holes (Fig. 9.1). The premature linking of the radial cracks will form a

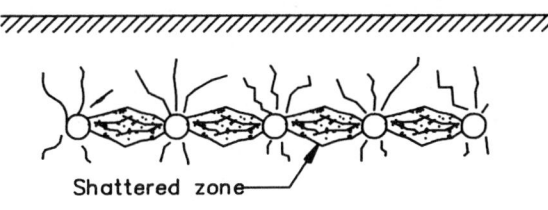

Shattered zone

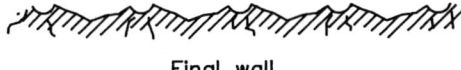

Final wall

Figure 9.1   Shattered zone from close spacing.

fracture plane between holes where gases can be vented prematurely to the atmosphere causing airblasts and flyrock. The venting will reduce the available useful work energy and in effect the holes will become overconfined since insufficient energy remains to properly break the rock in the burden. The overconfined condition will cause the ground vibration levels to increase. In spite of the close blasthole spacing, fragmentation of the burden rock will be poor. Conversely, if blastholes are spaced too far apart fragmentation will become coarse and rough walls will result (Fig. 9.2). Blastholes fired either simultaneously or delayed one from another will suffer if the spacings are too close or too far apart.

Blasthole spacing must be designed to overcome problems with burden stiffness. Therefore, when benches are low when compared to the burden, stiffness is a factor that must be considered. When benches are high, stiffness is no longer considered in our calculations.

There are two factors, therefore, that must be evaluated. We must determine if adjacent blastholes function either near simultaneously or delayed one from another. We must also determine if benches are classified as low or high as compared to the burden stiffness. Whether holes function simultaneously or delayed is easily determined by considering the periods of initiators used. To determine if benches are low or high must be tied to a physical dimension, such as the burden and bench height from which the stiffness ratio or L/B is calculated. If L/B is less than four and greater than one, benches are considered low and bench height must be considered in our calculations. On the other hand, if L/B is greater than four, bench height is not used in our calculations. There are, therefore, four separate conditions which must be discussed: instantaneous initiation on low benches, instantaneous initiation on high benches, delay initiation on low benches, and delay initiation on high benches.

Blasters in the field realized that adjustments to spacing had to be made based on timing and bench height. Drinker (1888) reported that in Bohemia as early as 1725 it was known that near simultaneous initiation required broader spacings than

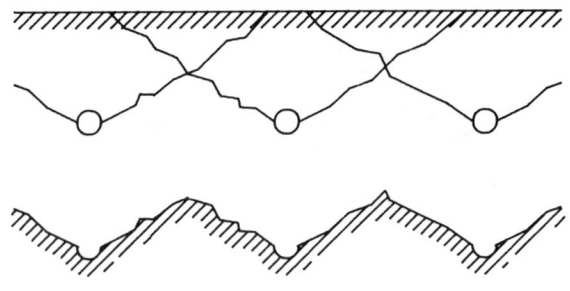

Final wall

Figure 9.2 Rough walls from excessive spacing.

when firing with delays. Gillette (1916) indicated that high benches required larger spacings then low benches. These observations were made in the field, however methods to determine the amount of compensation was not known. A research project was conducted by Konya (1968) to define the amount of compensation needed in spacing to overcome the effects of these two variables. The study was conducted in plexiglas, mortar, and dolomite. Small scale models were first used for the tests to define trends. Figure 9.3 is a graph which relates stiffness ratio L/B with the spacing ratio (spacing divided by burden). The data demonstrates that once bench heights were four times the burden (L/B = 4) or more no additional spacing compensation was needed. When L/B was less than four, spacing ratios were reduced as borehole length decreased. This same trend is indicated for both instantaneous and delay initiation. Field data was collected from strip mine blasts, quarries, and construction blasting operations and evaluated in the same manner as was previously done for the small scale model studies. Field data indicated similar trends (Fig. 9.4) with both delayed and instantaneous initiation between holes in a row. The spacing ratios were higher at the same stiffness ratio in the model studies than were used in the field. The difference in results can be attributed to the criteria used in the model studies to evaluate maximum spacing distance which could be achieved for a particular stiffness ratio. In small scale studies the criteria was whether or not shearing and displacement resulted between holes. If the rock material between blastholes totally sheared regardless of the fragmentation size, the results were considered positive. If the material between holes did not shear the

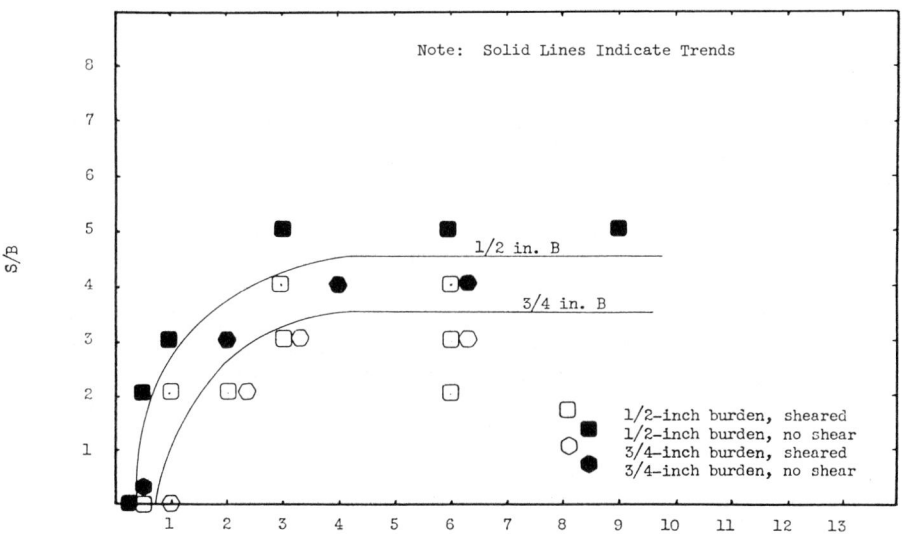

Figure 9.3 Explosive spacing/burden versus charge length/burden for mortar (instantaneous initiation).

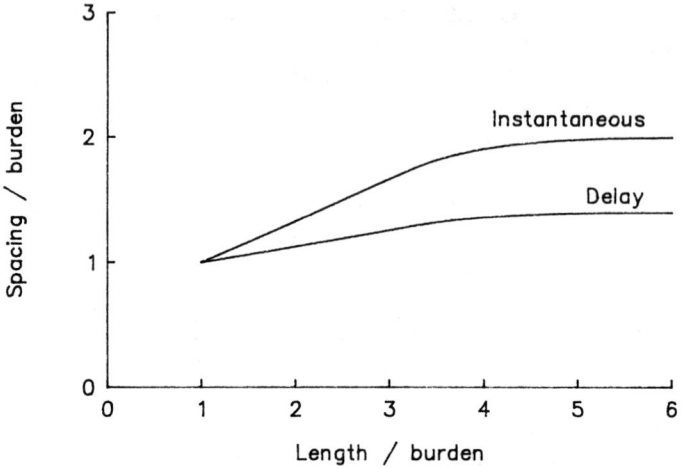

Figure 9.4 Explosive spacing/burden versus charge length/burden (field data).

results were considered negative. In this manner, a simple criteria was established to compare results. The field studies were evaluated on different criteria. The field studies used two criteria for evaluation, fragmentation, and shearing between blastholes. Fragmentation had to be within acceptable ranges for the results to be considered positive. The difference in evaluation criteria may account for the difference in magnitude between the field and laboratory results.

Both studies indicated that four different spacing ratios exist depending on timing and stiffness. For instantaneous initiation between holes in a row, a curvilinear relationship resulted when L/B was less than four. When L/B was greater than four, the relationship was linear. This same trend with different magnitudes also appeared for delayed initiation. Mathematical functions were obtained for these four different conditions. For example, in the laboratory study when L/B was less than four with instantaneous initiation, the relationship could be expressed as

$$S = 0.923 \, B \, \log_e (L/B - 1.17) + 2.4 \qquad (9.1)$$

Similar equations were derived for curvilinear functions in the field studies. These relationships, however, were cumbersome to use in the field, and therefore, simple mathematical functions were found which closely approximate these more complex equations. These relationships are given in Table 9.1.

The use of these relationships for pattern design are discussed in the following

TABLE 9.1 Spacing Equations

	L/B < 4	L/B ⩾ 4
Instantaneous	S = (L + 2B)/3	S = 2B
Delay	S = (L + 7B)/8	S = 1.4B

sections. Reasonable tolerance on blasthole spacings are normally considered to be within plus or minus 15% of the calculated values.

Instantaneous Initiation Low Benches

To determine the average spacing for the conditions where blastholes within a row fire near instantaneously and where benches are lower than four times the burden we can use the following equation: (See Table 9.1.)

$$S = (L + 2B)/3 \qquad (9.2)$$

where

$$S = \text{spacing in ft}$$
$$L = \text{bench height in ft}$$
$$B = \text{burden in ft}$$

We can use this equation to calculate the spacing for our design plan. When the proposed spacing is within plus or minus 15% of the calculated spacing, it is considered within reasonable tolerances. In no case should the spacing be less than the burden.

Example 9.1

Three-inch diameter blastholes, will be loaded with 2-inch diameter semigelatin dynamite cartridges. They will be fired row-by-row with instantaneous initiation along a row. The proposed pattern will be drilled with a 5-ft burden. The bench height on one portion of the excavation is 15 ft. Determine the spacing distance.
Check L/B for high or low bench

$$L/B = 15/5 = 3.0 \text{ (low bench)}$$

Check instantaneous or delay timing

Answer Instantaneous
therefore

$$S = (L + 2B)/3$$
$$S = (15 + 2 \times 5)/3$$
$$S = 8.33 \text{ ft}$$

The spacing is 8.33 ft. A reasonable tolerance is plus or minus 15%. The spacing, therefore, should not be greater than 9.5 ft or less than 7 ft. The spacing for the test shot would be at 8 ft.

Instantaneous Initiation High Benches

To function as a high bench, the bench height to burden ratio must be four or more. With instantaneous initiation between holes in a row, the following relationship can be used:

$$S = 2\,B \qquad (9.3)$$

where

$$S \; = \; \text{spacing in ft}$$
$$B \; = \; \text{burden in ft}$$

When using large diameter blastholes in mining operations, geologic structure becomes an important consideration since many pronounced joints can occur between blastholes. These structural features can cause fractures between holes to stop prematurely and produce less fracture than needed to provide optimum fragmentation. Blasthole spacing may have to be reduced to overcome structural features. In this case, Eq. 9.3 would represent the maximum spacing while actual spacing may be less in geologically complicated areas.

Example 9.2

The pattern in Ex. 9.1 is considered for a portion of the excavation where the bench height is planned to be 25 ft deep. What spacing should be used for the test blast on a 25 ft bench? Check L/B for high or low bench

$$L/B \; = \; 25/5 \; = \; 5 \; \text{(high bench)}$$

Check instantaneous or delay timing

Answer Instantaneous
therefore

$$S \; = \; 2B$$
$$S \; = \; 2 \times 5$$
$$S \; = \; 10 \text{ ft}$$

Delayed Initiation Low Benches

When the stiffness ratio is between one and four with delayed initiation between holes, the following relationship is used:

$$S = (L + 7B)/8 \tag{9.4}$$

where

$$S \; = \; \text{spacing in ft}$$
$$L \; = \; \text{bench height in ft}$$
$$B \; = \; \text{burden in ft}$$

Example 9.3

Four-inch diameter blastholes are bulk loaded with ANFO. The operator proposed to use an 8 by 12 ft drill pattern (8 ft burden and 12 ft spacing). Assuming the burden is correct, would the spacing be reasonable if the bench height is 12 ft and each hole is fired on a separate delay?
Check L/B for high or low bench

$$12/8 \; = \; 1.5 \; \text{(low bench)}$$

Check instantaneous or delay timing

Answer Delay
therefore

$$S = (L + 7B)/8$$
$$S = (12 + 7 \times 8)/8$$
$$S = 8.5 \text{ ft}$$

The proposed spacing of 12 ft is greater than plus or minus 15%. The proposed spacing is unacceptable.

Delayed Initiation High Benches

When the stiffness ratio is four or more and holes in a row are delayed, the following equation is used:

$$S = 1.4 \text{ B} \qquad\qquad\qquad (9.5)$$

where

$$S = \text{spacing in ft}$$
$$B = \text{burden in ft}$$

Example 9.4

The 8 by 12 ft pattern described in Ex. 9.3 is proposed for a section in the excavation where the bench height is 35 ft. Is the proposed spacing acceptable?
Check L/B for high or low bench

$$35/8 = 4.38$$

Check instantaneous or delay timing

Answer Delay
therefore

$$S = 1.4 \text{ B}$$
$$S = 1.4 (8) = 11.2 \text{ ft}$$

The proposed spacing of 12 ft is reasonable, since it is within the range of plus or minus 15% of the calculated value.

Pattern Construction

In order to maximize fragmentation and minimize unwanted side effects from blasting, the controlled variables such as burdens, stemming, subdrilling, spacing, and timing must be selected such that all variables are working together. To better understand the relationship between the variables, figures will be used to illustrate the effects of having properly matched variables and improperly matched variables. Unless otherwise specified, it will be assumed that there are no geologic complications and all bench heights are at least four times the burden.

When a blasting pattern is constructed, each and every hole must be analyzed to determine if it will respond properly. Analyzing spacings or drill burdens without consideration for initiation timing does not produce a true picture of what will occur when the blasthole is fired. There is often a difference between what the blaster considers as true burden and spacing and what the driller calls burden and spacing. Drillers are taught to call the distance between rows which are parallel to the face as burden. They are also taught to call spacing the distance between holes along a row parallel to the face. The blaster always considers the distance from the hole to the face as burden and this can be different than drill burden on multirow shots. The reason for the difference is that *relief* is defined as the distance to the open face or the internal face. The internal face is the face that can be created by holes firing on an earlier delay. The manner in which blastholes are timed can change a boreholes orientation to an internal face, therefore, true burden or resistance on that hole also changes. If the internal face is running at a 45° angle to the ledgeface, the spacing distance would also change since rows would be working towards the internal face and not the ledgeface.

If a pattern is properly designed, you will notice a repetitive sequence in the crater forms broken per hole. Different crater shapes will be created from independent holes firing depending upon the relationship between the blasthole and the free face. This can be seen in Fig. 9.5. To make analysis easy, one can assume that the

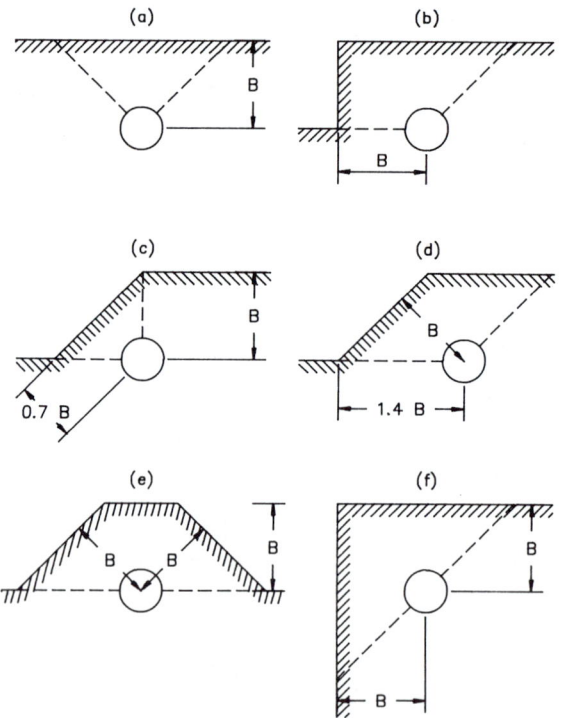

Figure 9.5 Crater forms which result from borehole orientation to face(s). (plan view)

breakage angle between the burden line and the edge of the crater is approximately 45°. If a blasthole has more than one burden direction at the time of its detonation, the distance to the free face along both burden directions should be equal. Figure 9.5(a) illustrates the breakage angle formed when one vertical free face is present. For the purposes of this analysis, the horizontal free face or the bench top will not be considered since from previous discussion it was evident that explosives preferentially function radially away from the blastholes. In Fig. 9.5(b) two free faces are present and form a 90° angle, breakage patterns would be different than in Fig. 9.5(a). In Fig. 9.5(c) a corner cut illustrates a different area of breakage because of the orientation of the face. If the blasthole is on a corner with two free faces, the breakage area is equivalent to two craters of area shown in Fig. 9.5(f). In Fig. 9.5(e), the crater formed will be larger than in any other geometry. The same amount of explosive is used in each blasthole in these examples; however, different volumes of rock are broken depending upon the orientation to the free face. This simple example shows why *powder factor*, the amount of explosive used per cubic yard, is not a constant number for a particular rock type, even if the rock type and explosive are identical to that used on other blasts. The crater forms produced actually control the amount of explosive used.

Figures 9.6 through 9.11 illustrate the effect of timing and spacing for a single row pattern.

Figure 9.6 represents a single row blast with progressive delays between holes. The spacing distance was drilled equal to the burden. The dotted lines on the

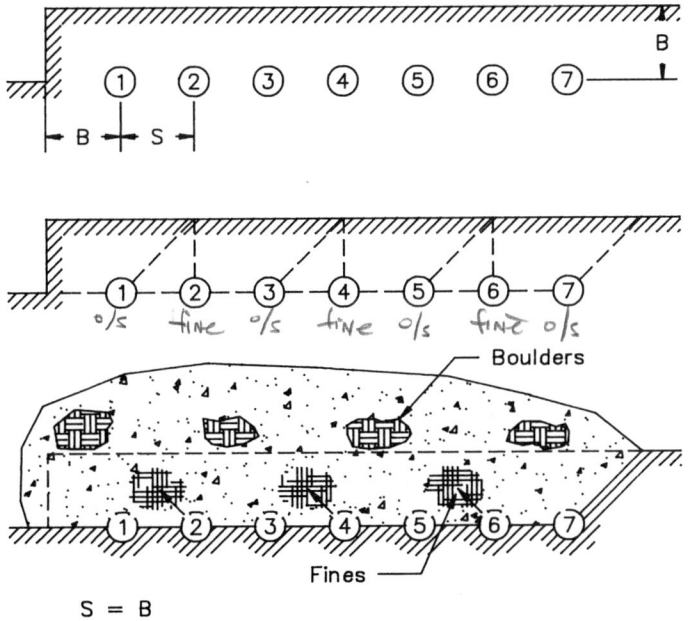

$$S = B$$

Figure 9.6 Spacing equal to burden, progressive delays.

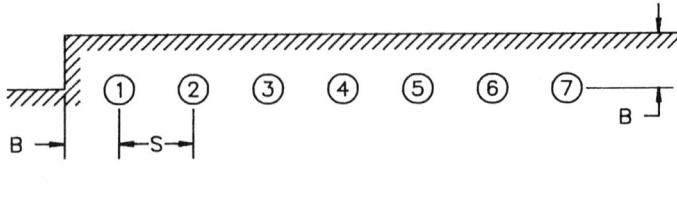

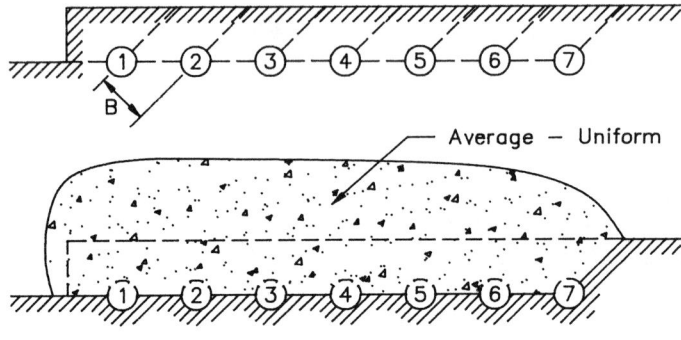

$$S = 1.4 \ B$$

Figure 9.7 Spacing equal to 1.4 burden, progressive delays.

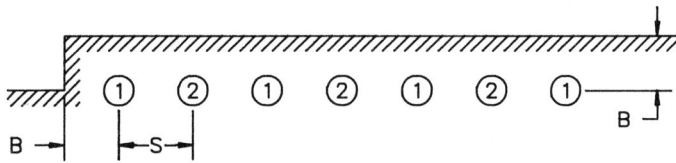

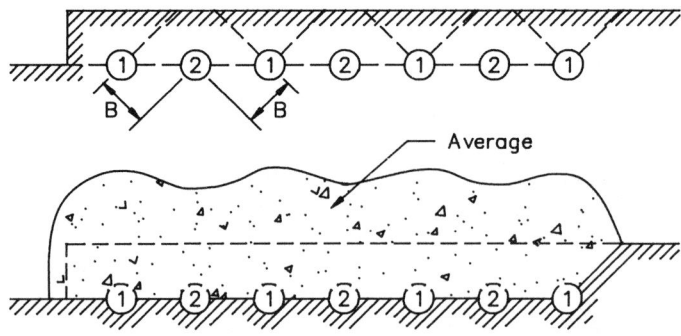

$$S = 1.4 \ B$$

Figure 9.8 Spacing equal to 1.4 burden, alternating delays.

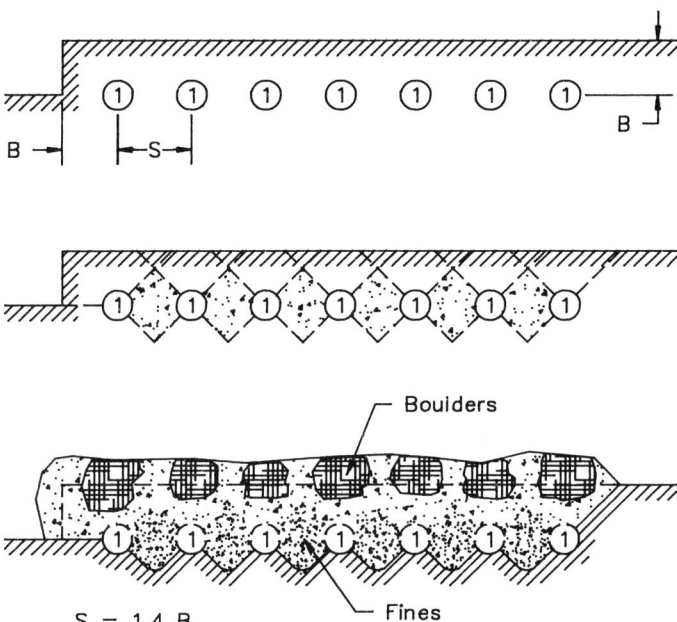

S = 1.4 B

Figure 9.9   Spacing equal to 1.4 burden, instantaneous firing along row.

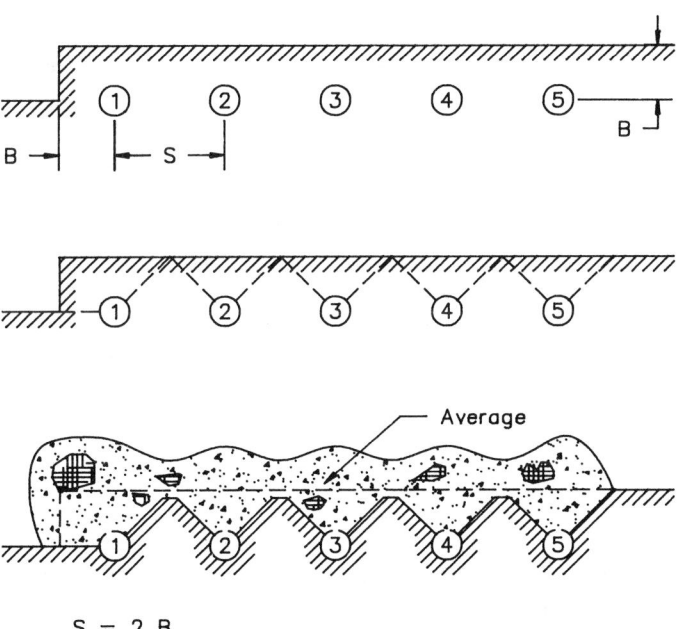

S = 2 B

Figure 9.10   Spacing equal to 2 burden, progressive delays.

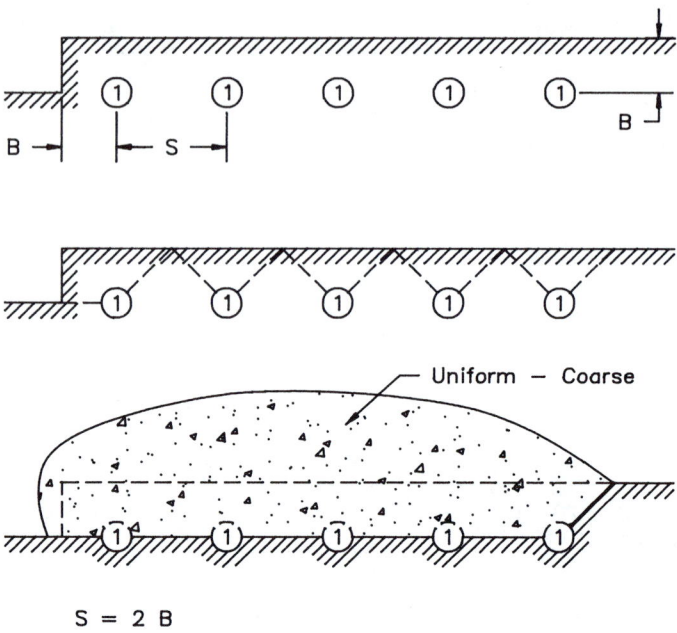

$$S = 2 B$$

Figure 9.11 Spacing equal to 2 burden, instantaneous firing along row.

figure indicate the expected breaklines for the crater formed from each individual hole. The radial crack pattern is totally developed before the adjacent hole fires when delay blasting is employed in the field. Figure 9.6 shows that two distinct patterns of breakage occur and two different crater forms are employed. If the spacing is equal to the burden the different crater forms will produce different size distributions of material. Holes numbered 2, 4, and 6 will produce finer material and move it further away. Holes numbered 1, 3, 5, and 7 will produce a coarser distribution with the possibility of a small boulder occurring off each corner.

Figure 9.7 represents the same pattern, the only difference in the pattern is that the spacing is extended to 1.4 burdens. Each blasthole breaks the same volume of material with exception of hole Number 1. Most uniform fragmentation results from one blasthole to another if the same volume of rock is broken with the same amount of explosive in the identical same fashion. This blast should produce an average uniform size material from hole-to-hole.

Figure 9.8 presents a blasting pattern drilled and loaded in the identical fashion as that in Fig. 9.7. The difference being that although holes are delayed, they are delayed in a different manner. Rather than having progressive delays along the row, alternating delays are used. Patterns such as this will produce different breakage from the Number 1 delays to the Number 2. The fragmentation will be different from the previous pattern. Although we have the same amount of explosive in every hole; the volume of rock broken and the manner in which it is broken is different from one hole to another. We would actually have three different size distributions since three different crater forms are used in this pattern.

Figure 9.9 presents a blasting pattern which is drilled and loaded identically to the previous one discussed. Different initiation timing is used in this example. All holes will fire near simultaneously. Although the energy density is the same in this pattern as in the others there will be reinforcements of energy between holes because of the initiation timing which can cause intense radial cracking and blowout along the line between the holes. When this premature energy loss occurs, burden rock can remain unbroken and form large boulders.

Figure 9.10 represents a blasting pattern which was drilled with spacings equal to twice the burden. The pattern is improperly delayed with progressive delays. The pattern has broad spacings which will only be broken by using simultaneous initiation along a row, however in our example, progressive delayed initiation is used. With no reinforcement of energy between holes, rock between holes will be improperly broken. The spacing distances is too large for the timing employed and holes begin to form individual craters.

Figure 9.11 has the same drill pattern as Fig. 9.10. The difference is that all holes are fired near simultaneously. A simultaneously fired row with broad spacings will break, however it will produce more uniform coarse fragmentation. The fragmentation from a pattern with holes fired near simultaneously will always produce coarser fragmentation than one that is properly drilled and timed for delay blasting.

Figure 9.12 is a diagram of the traditional box-cut or V-cut. You will notice on the diagram that the second and subsequent rows are further apart than the first. They are drilled at this distance to provide the same true burden on rows 2 and 3 as would appear on the first rows. Rows 2 and 3 are working to an internal face denoted by the dotted lines rather than the ledgeface. To maintain the proper geometry the extended spacings are used between the rows. Most holes in this pattern have the same spacing along the internal rows and is equal to twice the burden.

The holes most likely to cause blow-out and high vibration on the V-cut are holes on the longest delay or Number 6 delay. These corner holes are more heavily confined than other holes in the pattern. If the Number 5 holes do not fire at the proper time, Number 6 has no relief other than the surface and therefore will blowout. Rules of thumb have existed for decades which say "to avoid blow-out in back rows or corners, skip a delay in the corners." These rules of thumb may have been helpful under certain applications, but if the delays are improper, increasing the delay in the corners may still be insufficient to get the proper breakage action. A much better alternative would be to lay out the pattern in a different manner.

Figure 9.13 shows the traditional V-cut, but instead of blasting a rectangular pattern area, we will blast out a trapezoidal area. The pattern shown in Fig. 9.13 does not utilize the Number 5 and Number 6 holes. Experience has shown that not forcing the rock to break into a 90° corner will produce better walls and less chance for violence.

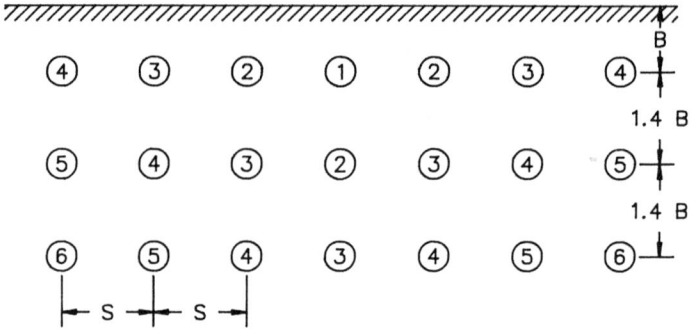

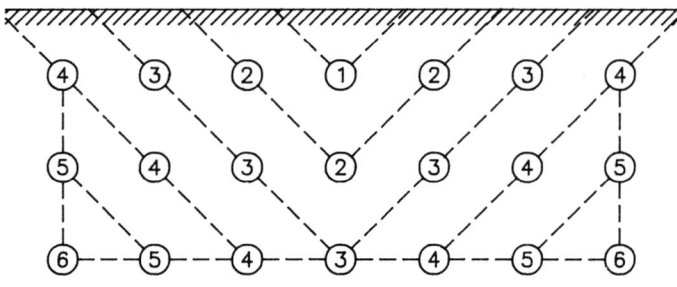

S = 1.4 B

Figure 9.12   V-cut (square corner), progressive delays, S = 1.4B.

Figure 9.14 is a box cut, which is delayed. No hole is firing at the same time as its adjacent neighbor. You will notice that the drill pattern is different than the previous V-cut pattern. Instead of rows being 1.4 burden distances apart, they are only a burden distance apart on the delayed box-cut. The reason for this is that most blast holes in the pattern actually sense 2 burdens which should be equal. They sense 2 burdens because they are working towards two internal faces.

Another difference in this pattern is that the holes on each end of the rows are pulled in from spacing distance 1.4*B* to burden distance. This is done to change the direction of the motion of the walls and reduce *endbreak*, breakage into the next unshot end area.

Figure 9.15 utilizes the same drill and loading pattern as the delayed box-cut. There is a difference, however, in initiation timing. Initiation timing on this cut utilizes alternating delays. Fragmentation from this pattern will be larger than that using the same drill pattern with progressive delays.

If an operator uses the V-cut, box-cut as indicated in Fig. 9.12, he may

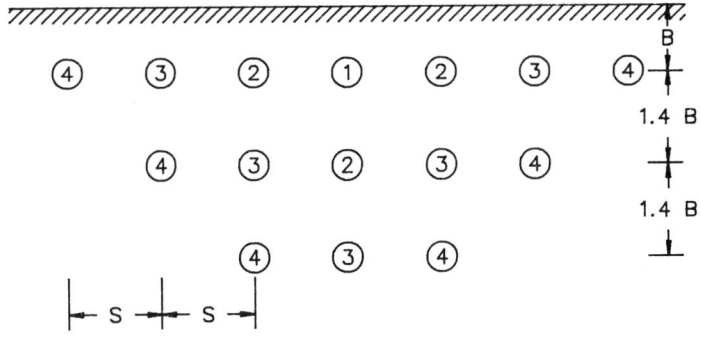

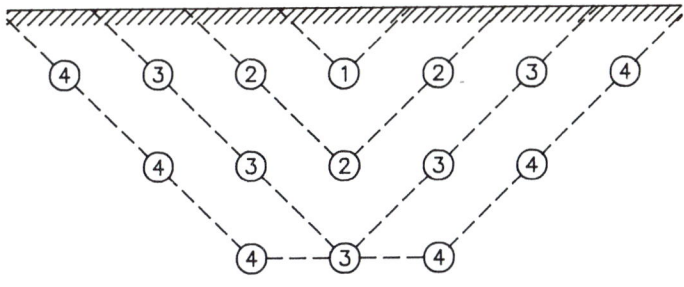

S = 1.4 B

Figure 9.13   V-cut (angle corner), progressive delays, S = 1.4B.

thereafter shoot corner cuts as indicated in Figure 9.16. The layout and timing of this pattern is identically the same as half of the box-cut shown in Fig. 9.12. This pattern, however, still offers potential problems with corner hole blowout.

If the V-cut shown in Fig. 9.13 is employed, leaving a trapezoidal pattern area, we could then use a corner cut as shown in Fig. 9.17 to further develop the bench. The last row of holes, the Number fours in Fig. 9.17, break at greater than a 90° angle in the corner, thereby causing less wall damage and a lower probability of blowout. The direction of rock motion from subsequent rows would be toward the dotted breakline shown on the diagram. Rock motion would be in the direction perpendicular to the dotted breaklines.

If the direction of rock motion and placement would change by 45° and move toward the vertical face as shown in Fig. 9.18, we would need to change the drill pattern to accommodate the change in direction of motion. If you compare the true burden on each blasthole in Fig. 9.17 and 9.18 you will notice that there is no difference. The spacing along rows on internal faces is also the same. Both patterns

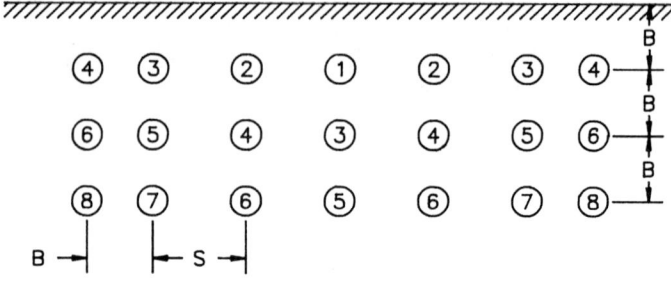

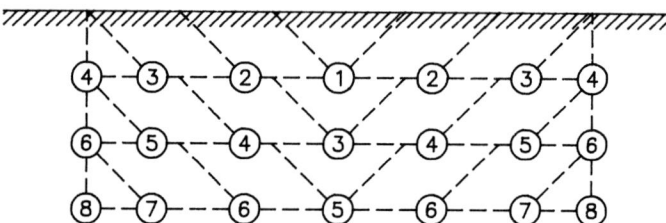

S = 1.4 B

Figure 9.14 Box cut, progressive delays, S = 1.4B.

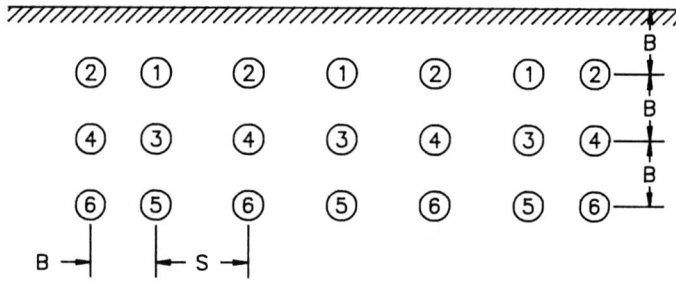

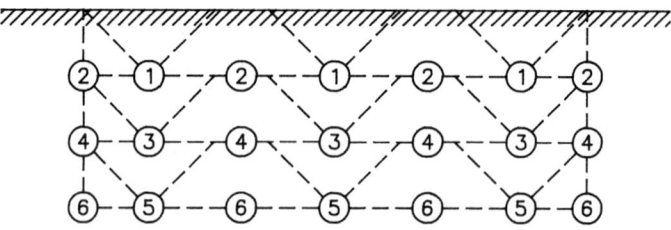

S = 1.4 B

Figure 9.15 Box cut with alternating delays, S = 1.4B.

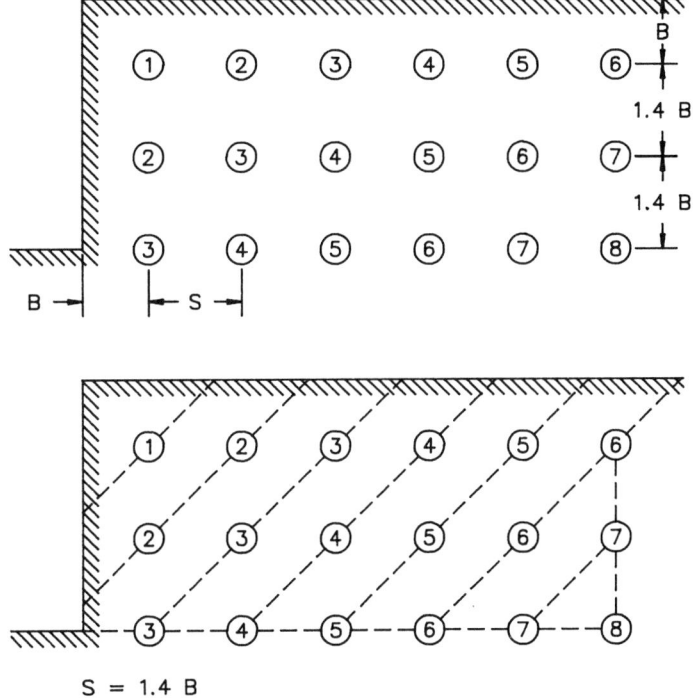

Figure 9.16   Corner cut fired on echelon, S = 1.4B.

fire in a manner where the second row of holes is staggered at the time of firing from the previous row. In other words, both of these patterns behave identically the same, however they place broken rock in a different location.

Very often it is difficult for the blaster to understand that in order to change the direction of the rock motion and have each hole maintain the same true burdens, spacing, and timing relationships, a different drill pattern must be used. Very commonly, blasters try to change direction of rock motion by changing only timing. They do not change the drill pattern. Therefore, the burden and spacing ratios within the pattern have changed, as a result of the change in timing. It is often incorrectly assumed that if the amount of explosive within the rock mass is not changed, the fragmentation results should be the same and only displacement direction is changed by changes in timing. This is incorrect because two factors control breakage, the amount of explosive used and the geometric relationship between holes which includes timing between holes.

If delayed patterns are used for the box-cut as earlier described in Figs. 9.14 and 9.15 we could use corner cuts as shown in Fig. 9.19. Each blasthole is breaking the same volume of rock in the same manner and therefore we would expect the most uniform fragmentation.

Experienced blasters in mining and construction realize that on short benches

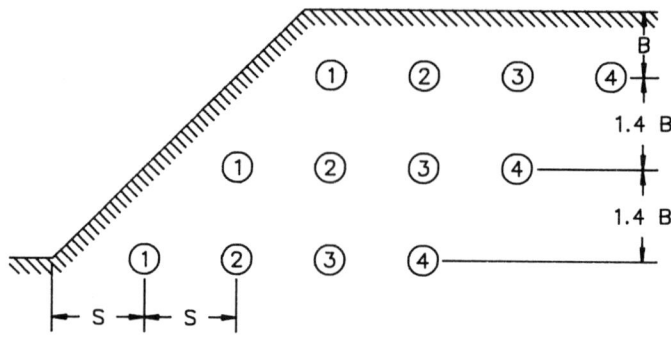

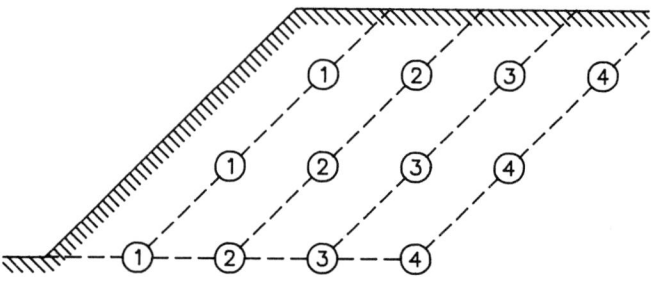

S = 1.4 B

Figure 9.17 Angled corner cut, fired on echelon, S = 1.4B.

an equilateral triangle drill pattern works better than a rectangular drill pattern. The reason it is called an equilateral can be seen in Figure 9.20. If we connect any three holes in any manner in the pattern, an equilateral triangle will result. If we analyze the pattern we understand why this pattern does a better job on low benches. When bench heights are approximately equal to the burden, the spacing is nearly equal to the burden (Fig. 9.4). Drilling an equilateral triangle pattern accomplishes that end result. Equilateral triangles will do a better job of breakage on low benches, but the breakage will not be nearly as fine as would result on higher benches with the proper pattern. Although this pattern somewhat compensates for the effects of stiffness, fragmentation results suffer.

Rip-Rap Production

Rip-rap is larger size rock normally used to protect banks or slopes from the effect of water and erosion. Rip-rap can weigh a few pounds or a few tons depending upon the end use of the product. Small size rip-rap can be produced in production blasts by increasing the burden distance and reducing the spacing distance.

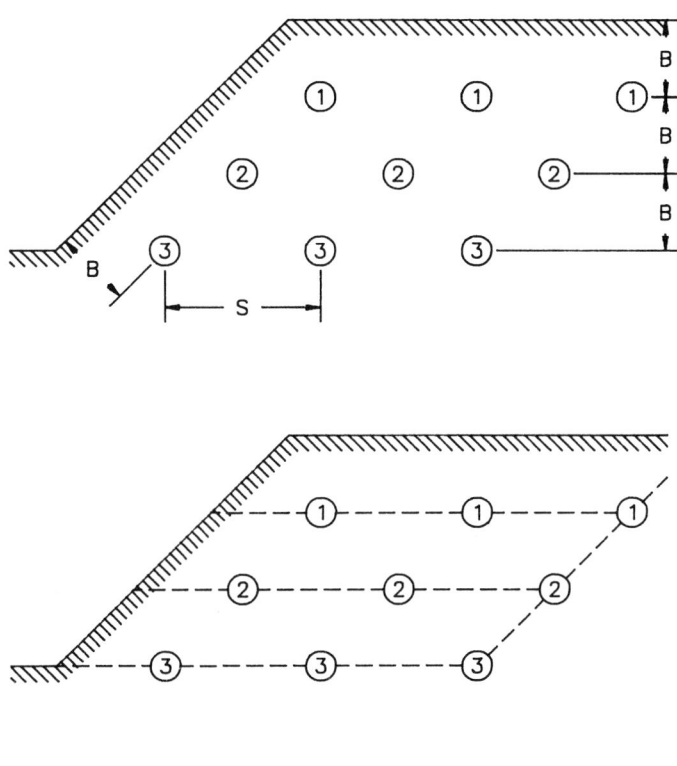

$$S = 2 B$$

Figure 9.18   Angled corner cut, fired instantaneously along rows, S = 2B.

Large size rip-rap, on the other hand, weighing thousands of pounds must be produced using a different technique. Large stone for breakwater walls must be undamaged so that the action of waves and freezing action will not deteriorate the rock prematurely. Extreme care must be taken to produce unfractured rock. This can be accomplished by using principles of controlled blasting along with the production blast. As an example, blastholes can be drilled with excessive burdens and minimum spacing. Blastholes are loaded lightly to prevent major damage from occurring around the borehole. When the blast is fired, large pieces of unfractured rock are produced (Figure 9.21). Not every rock mass can be used for rip-rap production. The rock must be either massive or interbedded with cohesion across the bedding planes.

Rock Piling Considerations

The function of the blasting pattern is not only to fracture the rock to the desired size distribution, but also to pile or place the rock in a manner that is most economic to handle in the next step of the operation. The type of equipment that will

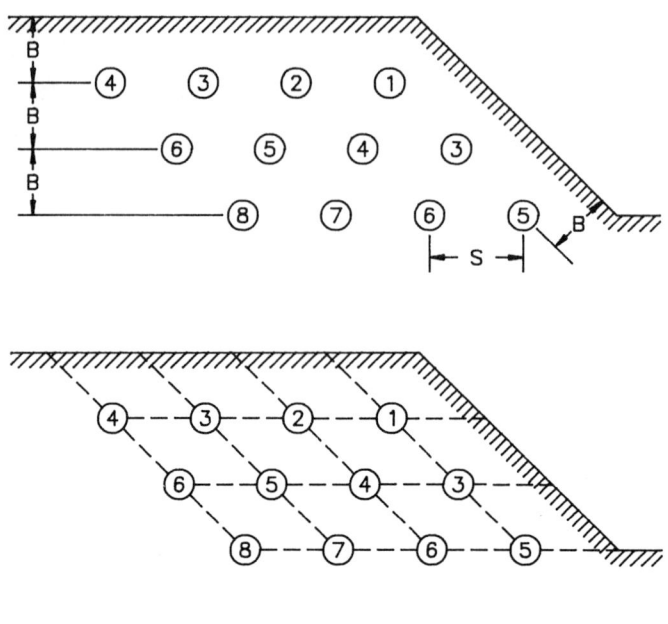

S = 1.4 B

Figure 9.19   Angled corner cut fired on progressive delays, S = 1.4B.

be used for digging the blasted material is, therefore, an important consideration when the blast is designed. If the benches are relatively low and a power shovel is used for loading, you want to stack the rock to ensure a high bucket-fill factor. On the other hand, if benches are high and an end loader is used for digging, intentional scattering of the broken rock is desirable. To ensure the proper piling of broken material, the following principles should be considered in the design process:

1. Rock movement will be parallel to the true burden dimension.
2. Instantaneous initiation along a row causes more displacement than delayed initiation.
3. Shots delayed row-by-row scatter the rock more than shots arranged in a V-cut.
4. Shots designed in a V-cut produce maximum piling close to the face.

Control of Ground Vibration, Airblast, and Flyrock

The methods of pattern construction previously discussed have shown the general timing sequence for blastholes. The actual time in milliseconds which would be used in these patterns will also control scatter or piling along with airblast, flyrock, and ground vibration. The general guidelines for producing proper timing

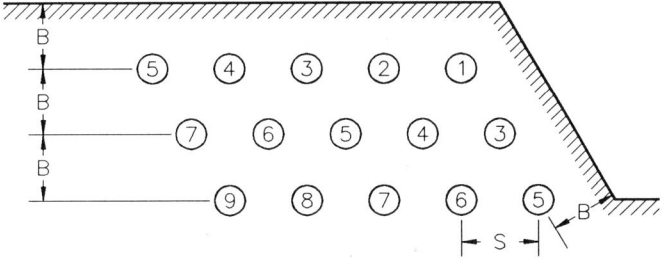

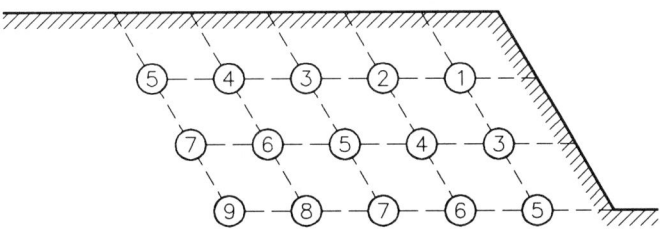

Figure 9.20   Equilateral triangle pattern for low benches, S = 1.15B.

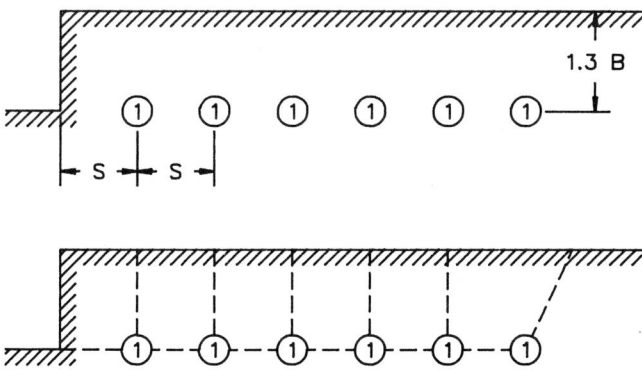

S = B

Figure 9.21   Pattern for rip-rap production S = B.

were given in Chap. 8. These must be considered in the selection of the actual timing in milliseconds both hole-to-hole in a row or row-to-row for the patterns described in the previous section. The combination, therefore, of blasting pattern sequencing along with actual timing further controls scatter or heaping of the pile.

Sinking cuts.   When starting off from a flat rock surface and dropping down to a lower level, such as in highway construction, foundation placement, or blasting for a bridge pier, a blasting pattern commonly called a sinking-cut, drop-shot, or drop-cut will be used. This shot is different than the other production blasting patterns previously discussed because at the time the shot is initiated there is only one free face and that is the horizontal top surface of the rock.

The first holes to fire in this type of pattern function totally different than those previously discussed. These opening holes must create the second free face toward which the rock from later delayed holes can push, bend, or move. Timing of these holes is critical, since too short a time delay between the initiation of the first or center holes and the subsequent holes cause poor breakage along with extreme violence. Figure 9.22(a) indicates the sequencing from row-to-row by increments of only one cap period. The pattern in Figure 9.22(b) shows a totally different firing sequence, which allows additional time for movement before each subsequent delay fires. The pattern in Fig. 9.22(a) also has many holes firing on the same delay period, which increases the ground vibration level. Vibration from this type of shot will be higher than from other production rounds because the first holes to fire are heavily confined at the time they detonate.

To better understand the functioning of a sinking-cut, the pattern in Fig. 9.22(b) will be discussed in detail. The pattern in Fig. 9.22(b) has only four holes firing per delay period. This is an important difference between the two patterns, especially near the center of the shot. If too much rock moves toward the center of the shot at one time, the center of the pattern may pack and not move. If this occurs, the remainder of the holes in the pattern will rifle because they have no place to move laterally.

The Number 1 holes or the first holes to fire in the pattern in Fig. 9.22(b) are functioning differently than the rest of the holes in the pattern. For example, the Number 1 holes are all stressing area A as indicated in the diagram with a tremendous concentration of energy within the zone. Holes Number 2 and others thereafter use half the number of holes and approximately half the explosive to break a similar volume of rock. Holes marked Number 1 radially crack the rock, but cannot bend or displace it since there is no place for this type of motion to occur. Instead, the radial cracks are pressurized by the gasses and begin to lift upward as in a cratering shot. Holes Number 2, however, function differently because they are moving toward the area vacated by the Number 1 holes. Number 2 holes function toward a free face outlined by the break line of the Number 1 hole. They, therefore, radially crack and displace into the crater produced by Number 1 holes. Subsequent holes in the shot all have a vertical free face to work toward as did holes Number 2.

The pattern in Fig. 9.22(b) is somewhat different from other patterns pre-

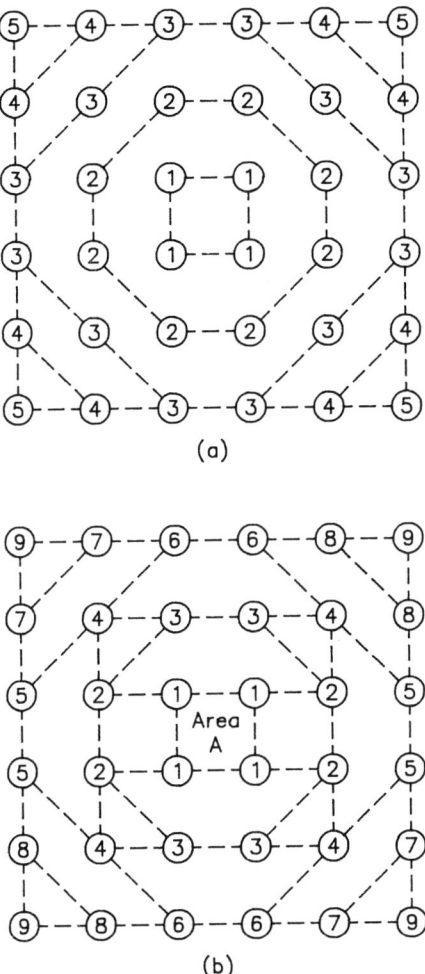

Figure 9.22 Sinking cuts, square pattern,
S = B.

viously discussed because the physical direction of the burden changes with each hole firing. If the pattern is laid out such that the top of the page is north, holes Number 2 sense a burden in an east–west direction where holes Number 3 sense a burden in a north–south direction. The burden is the most important dimension in a blast. To ensure that all holes have the same maximum distance as the burden, the pattern will be drilled square with a burden and spacing equal.

The Number 1 holes must break to grade to ensure that the subsequent holes can break to grade. If holes Number 1 break only partially to the grade line, the entire bottom of the shot will be high and above grade level. To ensure that Number 1 holes break properly, they should be drilled deeper than those in the remainder of the shot. The Number 1 holes should be subdrilled approximately twice as deep below grade as others in the blast.

Number 1 holes function differently than the remainder of the holes in the shot and are designed to crater. To control flyrock from the holes, it is a common practice that the Number 1 holes contain more stemming than the others. The stemming is commonly equal to the burden distance. The remainder of the holes will be stemmed to a depth of approximately 0.7 burden.

The final dimension which needs to be considered in a sinking-cut is the depth of the shot. Breaking to great depth is not easily accomplished. Gravity effects cause problems with rock motion necessary to produce the desired relief.

There are two rules of thumb which are considered when designing sinking cuts. The first states that the depth of holes should not be greater than half the lateral dimension of the pattern. This is to say that the cut depth will be one-half the distance obtained if spacing between blastholes in a row are added together. As an example, if the pattern width was 60 ft, then this rule of thumb indicates that the depth of the cut should be no more than half that, or 30 ft. A second rule of thumb states that the maximum L/B or bench height to burden ratio for a sinking-cut to function properly should not be greater than 4. For example, if the burden between holes in the pattern is 5 ft, a practical sinking cut depth of 20 ft would be realistic. On the other hand, if 6 1/2 in. holes were being used for sinking cuts with burdens of 15 ft, then practical depth of the cut might be as great as 60 ft. The greater the depth in a sinking cut, the greater the probability that the cut will not function properly and will not break totally to grade. Laminated rock with closely spaced bedding planes is more forgiving to errors in judgment than massive rock. When blasting massive rock, these ratios should be followed, while in laminated rock additional depth is often obtained.

A sinking-cut is meant to break and heap the rock with minimal cratering and rock ejection (Fig. 9.23).

To summarize the methods of finding the average dimensions for a sinking-cut we must separate what is done on the opening holes (Number 1) and all the others. The burden for this pattern is calculated as described in Chap. 7.

General Pattern Dimensions	Opening Holes	All Other Holes
$S = B$	$T = B$ (chips)	$T = 0.7B$
$L = < 4B$	$J = 0.5B$	$J = 0.3B$
$L = \geq B$		

Hillside cuts.   Hillside cuts can be difficult to control, since in most instances the rock cannot be thrown from the hillside. If the purpose of the pattern was to scatter the rock down the hillside, there would be no problem is designing the blast. When it is the intent of the operator to keep as much rock as possible in the cut itself, procedures can be used that are either similar to a modified sinking-cut or similar to a modified V-cut. The method of timing of the blastholes will ensure rock

Figure 9.23   Blasting a sinking cut in limestone. (Courtesy of IRECO Inc.)

movement in a manner to keep the rock pushing toward the bank rather than pushing toward the slope. An example of this type of cut is given in Figs. 9.24 and 9.25.

On steeply sloping hillsides, the outer row of holes has very little depth. To produce the proper fragmentation, displacement, and piling, especially in massive rock, the operator must consider the general principles of rock breakage as described in Chap. 2. The L/B ratio must never be less than one. If large diameter holes are used with high benches, then it is often necessary to reduce blasthole size, burdens, and spacings on the outer edges of the slope where benches are low to maintain an L/B ratio greater than one. Air track drills with small diameter bits may be necessary to produce the proper results on the outer edges of the slope.

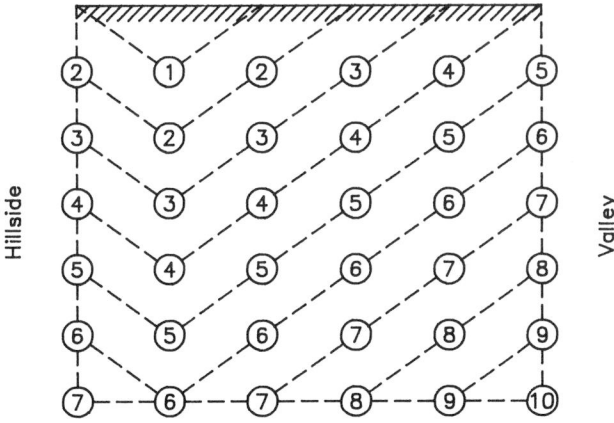

Figure 9.24   Hillside cut, S = 1.4B.

Figure 9.25   Hillside cut in shale.

Cast blasting.   Cast blasting or overburden casting is employed when it is desired to deliberately move the broken rock further than would normally occur on the average blasting operation. Cast blasting is commonly used in coal stripping operations since it is often more economical to use the explosive energy to move the overburden instead of the mechanical energy from power shovels, draglines or other excavating equipment. Cast blasting can increase the profit margin on the coal produced. It can also increase coal reserves by reducing the production cost on deep reserves. Cast blasting can significantly reduce cost. As one example, a coal mining operation increased their profits by an additional 30% of the sale price of the coal by using cast blasting. Cast blasting, however, is not only used in coal mining but finds application in some areas of rock excavation and to some degree in quarrying where either waste material or rock is deliberately scattered to reduce equipment utilization costs. For cast blasting, it is important that the burdens be measured accurately since an increase in burden, especially in front rows will slow motion and cause blastholes to rifle. When situations occur where the burden is not uniformed on the front rows, additional higher energy explosive in the larger burden zones is necessary to control rock velocity. Chiappetta (1983) proposed a first approximation of velocity from data he obtained from high-speed photography in dolomite granite and iron ore. The equation is as follows:

$$V_o = K \, (B/(\text{Energy})^{1/3})^{-\beta} \qquad (9.6)$$

where

$$
\begin{aligned}
V_o &= \text{Initial velocity in ft per sec} \\
K &= \text{site constant (25 as an example)}
\end{aligned}
$$

$$B \quad = \text{ burden in ft}$$
$$\text{Energy} = \text{ kilocalories per ft of explosive column}$$
$$-\beta \quad = \text{ site constant } (1.17 \text{ proposed})$$

This relationship will give a first approximation. The equation neglects the effects of burden stiffness. Burden and spacing ratios are also not considered in the equation.

GENERAL GUIDELINES FOR CAST BLASTING

1. Burden equals spacing with simultaneous initiation along rows.
2. Stemming equals burden.
3. Bench height should be approximately 4 times the burden.
4. Delays between rows should be between 7 and 14 ms per ft of burden.
5. Low-period delay caps should be used in the holes to minimize cap scatter and out of sequence shooting.
6. Short period downhole delays should be used so that adjacent holes do not cut off initiator leads from ground swell.
7. When possible, entire rows should be fired on the same delay.
8. If vibration control is necessary, short period delays, 17 to 25 ms should be used between holes in a row.

Utility trench design.   There are many considerations when designing a utility trench. The trench width is controlled by the size of pipe or utility which will go into the trench. You do not want to blast a 6-ft wide trench if only an 8-in. line is going into the ground. The size of the excavation equipment bucket which will clean out the trench must also be considered. In no instance can we design a trenching pattern, which has a width less than that of the excavator bucket.

Trenches are located near the surface of the earth, where you can encounter the most weathered, unstable type of rock. Often there has been decomposition of the rock resulting in clay or mud pockets and seams within the rock mass. The overburden, whether it be weathered rock or soil, may not be flat lying and this is an important consideration when the holes are loaded. One does not place explosives in the overburden above the solid rock. The blaster must know the actual depth to rock within each hole. To blast efficiently, explosives would be loaded in the hole and stemming must be placed within the rock itself, not just in the overburden.

Techniques which are used in bedded weak rock may not function well in solid massive material in utility trench blasting. Bedding planes will allow gas migration into the rock mass allowing more cratering action. On the other hand, similar techniques used in massive rock may not cause cratering. Instead, blastholes may rifle with little, if any, resulting breakage. In the following discussion, the difference in blasting technique between massive, hard materials, and interbedded, weaker rock will be reviewed.

If a narrow trench is desired in an interbedded rock mass, you can often use a single row of holes down the center line of the trench. The burden distance or spacing along the single row of holes would be calculated by Eq. 7.5. A minimum L/B ratio of 1 should be used in all types of blasting.

If the trench is to be shallow, smaller diameter holes will be needed than if the trench is to be deep. The timing should be such that holes will sequence down the row. If blastholes are all fired instantaneously, rock will be scattered in the nearby area. As bench heights are reduced, the probability of flyrock will increase and blasting mats may be necessary to control throw. The single row technique may not break properly in massive hard rock. Blastholes may rifle with little, if any, breakage between holes. In massive hard material, a double-row trench is commonly used.

The double-row trench is designed as indicated in Fig. 9.26. In massive materials, the blastholes should be placed at the excavation limit. In bedded weaker materials, it is often recommended that the blastholes be placed about a foot within the excavation limit since overbreak usually results. Placing the blastholes within a foot of the excavation limit, in massive materials, will produce poor results.

To determine if a utility trench pattern is within reasonable limits, the following guidelines are used:

1. The burden distance should be calculated by Eq. 7.5 and that burden is placed at the location indicated in Fig. 9.26. You will notice that this is not the actual

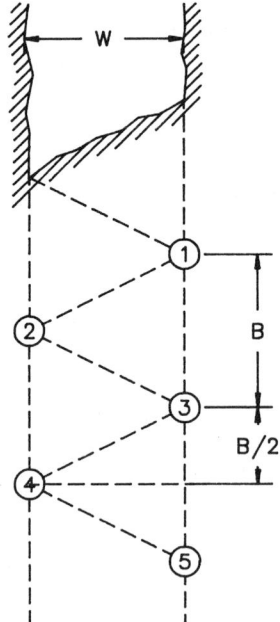

Figure 9.26   Two-row trench.

true burden since it is not the perpendicular distance from the hole to the face at the time the hole detonates. By placing the design burden as indicated in the figure will actual cause the true burden to be smaller. This is done to help compensate for the tight crater angle (approximately 60°) which requires more energy to break (Fig. 9.27).

2. The width of the trench must be between $0.75B$ and $1.25B$. If trench widths must be less than $0.75B$, then smaller holes and smaller powder charges should be used with burdens that are appropriate for these smaller charges. If trench widths must be greater than $1.25B$, either a larger borehole would be needed with its appropriate burden, or a three-row trench as indicated in Fig. 9.28 could be used.

3. The L/B ratio should be greater than 1.

Secondary blasting.   Secondary blasting is used when large boulders result from the primary blast. There are three common secondary blasting techniques used: mud capping, blockholing, and air cushion blasting.

Mud capping (boulder busting).   Mud capping or plaster shooting was previously discussed under the section on shock energy in Chap. 2. Mud capping utilizes an external charge placed on top of the boulder with a cap of mud placed on top of the charge. When mud capping is used, charges of between 0.5 to 1.0 lb of explosive per cu yd of boulder are normally sufficient.

Figure 9.27   Test blast of pattern for two-row trench.

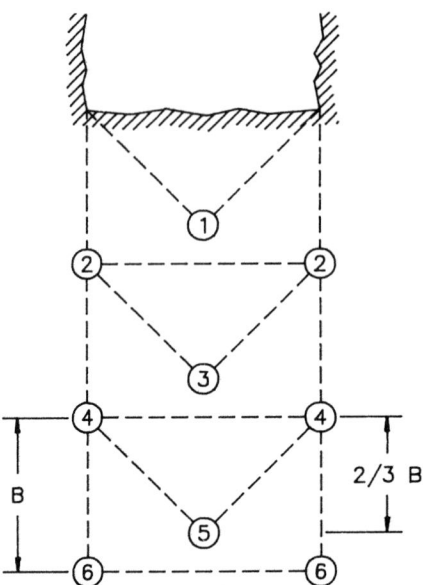

Figure 9.28   Three-row trench design.

Blockholing (boulder busting).   Blockholing simply means placing a hole or holes into the boulder and lightly loading these holes with explosives. The load is approximately 2 oz per cu yd for the test shot, and thereafter either increased or decreased depending on the type of rock being blasted. If the boulder is not square in shape and instead is rectangular, many small holes may have to be drilled and the powder load distributed between those smaller holes. Blockholing techniques utilize much less explosive than mud capping; however, the degree of fragmentation and the direction which the fragments fly is not controllable by the blaster, since the charges are functioning as cratering charges and breaking randomly in the direction of least resistance.

Air cushion blasting.   A technique similar to blockholing called air cushion blasting provides some control over the number of fragments and the direction in which the fragments fly. Air cushion blasting works as indicated in Fig. 9.29. A blasthole is drilled between 2/3 and 3/4 of the distance through the boulder. A charge which equals 2 oz of explosive per cu yd of rock is used for the test shot. Blastholes are stemmed to a minimum of 1/3 the depth of the hole. Common stemming materials are clay rather than crushed stone. The reason clay is used rather than crushed stone is that crushed stone must have distance to move up in the hole and lock into place to function properly. In general, with air cushion blasting the stemming distance is not long enough to allow the material to move and lock into place, therefore, clay is used, which will not lock into the hole but will provide a time lag between the time the hole is pressurized and the clay is ejected. The

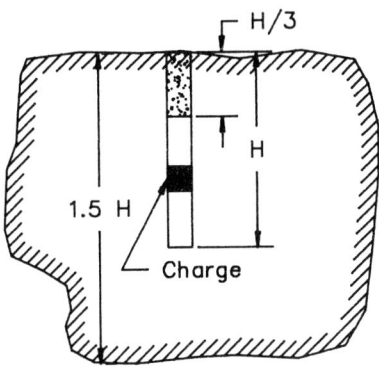

Figure 9.29   Air cushion blasting of boulders.

minimum depth of stemming with this type of blasting should be approximately 12 in. If stemming depths are less, holes may rifle, and little breakage will result.

When the minimum amount of stemming is used, the maximum air cushion occurs. The rock will break into the minimum number of pieces. Often, in massive materials an operator can predict with fair accuracy whether the rock will break into two or three large pieces, or four or five pieces. When air cushion techniques are used, little flyrock usually occurs with the rock normally remaining near its original location with little, if any, throw. If more fragments are desired, the air cushion can be reduced by increasing the amount of stemming in the hole. The more stemming placed into the blasthole, the more fragments will result and more violence will occur.

TROUBLE-SHOOTING BLASTING PATTERNS

For proper blast design, one must make the assumption that all blastholes release nearly ideal energy. In many types of blasting operations, blastholes release very little useful energy. These blastholes are called malfunctioning blastholes. Recent studies indicate that as many as 30% of all blastholes are malfunctioning. These malfunctioning blastholes are not an unusual occurrence.

We will discuss the cause of malfunctioning blastholes and methods to minimize the detrimental effect on both safety and cost.

Blastholes sometimes fire with excessive violence, while at other times, they may not fire at all. It is not uncommon to find unfired explosive in the muck pile. Normally, if this occurs, it is assumed that the misfire probably resulted from a cutoff or a missed hole and the incident is forgotten, until it reoccurs.

Originally a *"misfire"* meant that an explosive charge did not fire. *"Cut off"* is a general term used to either describe a cutting of the initiator leads before it has a chance to energize, or cutting off the powder column by shifting rock beds. *Missed holes* are those where the initiator was not connected into the firing circuit.

Whenever any amount of explosive is found after the shot, it is normally concluded that some type of cutoff occurred and that the cutoff was the reason for the unshot explosive. The blaster is normally confident that all initiators were properly connected into the circuit, therefore, it is assumed that the problem was not a missed hole. Shifting rock beds can cause cutoff powder columns and initiators in very seamy rock, where bed movement or movement through seams can more easily occur. Blasters often blame cutoffs as the problem for misfires in monolithic granite, limestone, or sandstone formations. Can this physically happen?

Misfires and cutoffs have been blamed for many other problems that result from improper behavior of a blasthole. Misfires or cutoffs have become generic terms meaning something did not go as planned and either no explosive detonation or partial or low-order detonation occurred.

There are many reasons why explosives do not detonate or partially detonate, releasing only a fraction of their potential energy. Blasters sometimes believe that high air-blast levels and high-vibration levels on a blast prove efficient energy release, in fact, it can be just the opposite. On some blasts, 30 to 50% of the holes do not properly detonate. Diagnostic tools available in the industry have proven conclusively that these problems are occurring.

Our general acceptance of misfires and cutoffs have led many operators to the conclusion that misfires and cutoffs are acts of God, which are normally caused by unusual geologic conditions and that they may not happen on the next shot. Therefore, it is often concluded that it isn't worth the time and effort to try and determine exactly what went wrong. If the actual problem could be found and corrected, the additional energy saved is available for fragmentation.

Malfunctioning Blastholes

Only those holes that did not shoot should be called misfires, and those that never received the detonation signal to fire are missed holes. Where blastholes did not function as intended and released little useful energy, they should be called malfunctioning blastholes. Malfunctioning holes result from many different causes (Fig. 9.30). The causes can be broken down into four general categories.

1. Insufficient energy release.
2. Initiator error or incompatibility.
3. Poor execution of the blasting plan.
4. Unusual local geology.

These four causes affect the energy release, breakage, violence, and create unpleasant environmental effects.

Insufficient Energy Release

Insufficient energy release occurs for many reasons. Field mixed explosives such as ANFO and ANFO–slurry blends, which are bulk loaded, can have a highly variable energy release depending on the composition of the mixture. The bulk

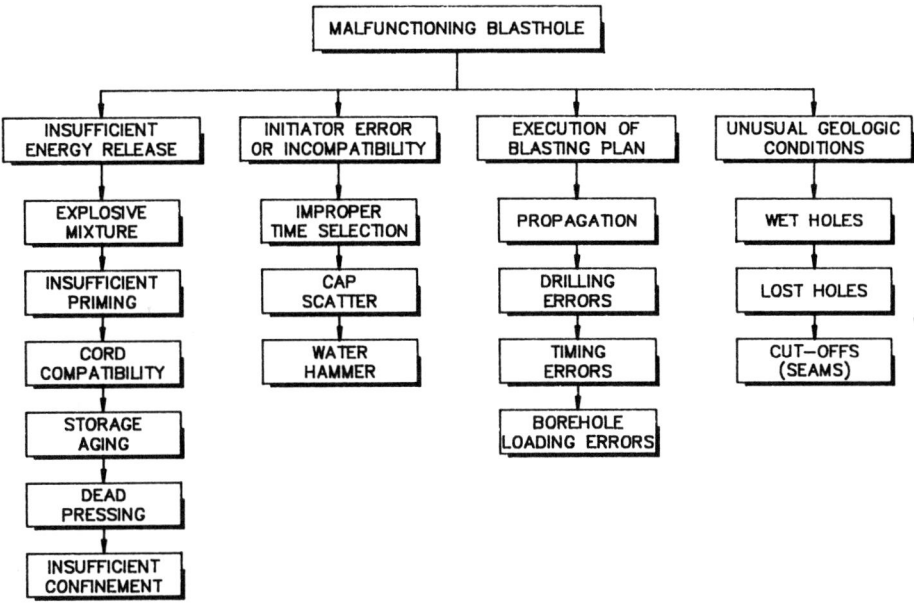

Figure 9.30 Causes of malfunctioning blastholes.

mixed explosives need to be near oxygen balanced to produce maximum energy just as in cartridged explosives. For ANFO alone, the most efficient fuel oil content is 5.7% by weight. It is common to find ANFO with fuel oil content varying between 2% and 10%. Field mixed explosives are more prone to variations in mixtures because of human error and less stringent quality control procedure.

The behavior of the explosive is controlled by the local environment in the blasthole and the effect of outside influences from the adjacent blastholes. A loss of confinement can be caused by holes shooting out of sequence and is a common example of an outside influence from an adjacent blasthole. Many factors can cause nonideal detonations. Blastholes are normally delayed from one another. Since all do not fire simultaneously, adjacent holes are effected by high-stress levels from earlier firing holes. The high-stress levels in some cases can cause precompression or dead pressing of the main charge. Precompression can cause some types of explosives to fire inefficiently and release only a small fraction of their available energy. High-stress levels can cause some explosives to propagate from deck-to-deck or even from hole-to-hole within the pattern. In the case of propagation, blastholes fire out of sequence and will override initiator delays.

We expect explosive products to release their rated energy during detonation. The energy level of products of the same type can vary from one batch to another. Extended storage of the explosive after manufacture can drastically change its characteristics and have effects on its response to local environmental conditions.

Water conditions in blastholes range from totally dry holes to those which not only have water clear to the collar, but actually have water under pressure which causes an artesian flow from the collar of the hole. Water conditions will influence the type of explosive used and the cost of the project. Project costs rise as water conditions get worse. Hence, more water resistant explosives will be needed. Many different kinds of problems can be encountered because of water conditions. The most serious problem results when powder has insufficient water resistance and becomes wet and will not function properly. A few examples of water problems which result in low-energy yield are given below.

1. The blaster did not take into account when loading the hole that the water level would rise as some of the explosive displaces water in the hole. Therefore, he used cartridged powder in the lower portion of holes but did not totally load out of the rising water. When bulk powders were subsequently used, water invaded the powder causing it to react inefficiently producing low-energy yields.

2. At the time of loading, cartridge powder may occupy the portion of the hole under water and dry bulk ANFO may be placed in the portion of the hole which is dry. A blast may take many hours to load and water can rise into the bulk powder.

3. In many patterns, some holes are wet to the collar and others are totally dry. What often happens is that the blaster does not realize that by placing cartridge explosives in holes that contain some water, the water level will rise. If there is communication between the wet hole at a higher level and an adjacent dry hole, water from the wet hole can be forced up and through the fracture system into what was previously loaded as a totally dry hole, thereby causing problems.

4. Some explosives produce low-order detonations under high-water head as a result of precompression from the water pressure.

5. Water flowing through blastholes can cause serious problems. Some operators consider bulk loaded water gel and emulsion explosives to be waterproof. They are only waterproof or water resistant in stagnant water. Flowing water will cut through bulk emulsion of water gel rendering that portion of the column useless and unable to transfer detonation from one section of the column to another. If flowing water is encountered, cartridge explosives must be used to ensure that the explosive column will not be cut by flowing water dissolving the explosive.

Initiators Error or Incompatibility

Choosing the proper initiator is critical to proper blasting performance. Too often, blasters are more concerned with price per unit and number of periods and do not pay sufficient attention to the conditions of intended use.

Initiators are placed in the blasthole to transfer the detonation signal from one hole to another at a precise time. The initiator is subject to high stresses from adjacent holes. Whenever blastholes are used in close proximity to one another, initiators can be damaged or destroyed by the firing of adjacent holes. This destruction may cause the initiator to fire improperly or may cause the initiator to totally misfire. This phenomenon is commonly called the *water hammer effect* (Fig. 9.31). Some initiators better withstand the effects of water hammer than others, and in some applications this is a major concern when purchasing initiators.

Detonating cord downlines that are too energetic can cause large energy losses in the powder column. In other applications, the cord has caused the explosive to prematurely burn, releasing little energy. When explosives are prematurely initiated by detonating cord downlines, the cord acts as an inefficient primer. The charge may deflagrate or go into a low-order detonation causing a large energy loss.

The selection of initiators simply for sequencing holes is a common error. The precise number of milliseconds of time between initiator periods is important in rock breakage. Initiator firing times are not precise and some initiators have large timing errors, both in nominal firing times and in cap scatter. In order for blastholes to properly function, the design of the pattern must consider both the true (nominal) firing time and cap scatter. If this is not done, holes commonly fire out of sequence. Methods exist to design patterns that will function properly in spite of the initiator inaccuracy. Poor timing is a common occurrence which causes holes to malfunction. As an example, if holes fire either out of sequence or without the proper time window from one hole to another, fractures will not move in the directions planned. Instead, the cracks can extend into adjacent holes. Figure 9.32 shows what can happen if holes fire out of sequence. Delay Number 4 has fired before Number 3. Adjacent holes can suffer a loss of confinement or a disruption in the powder

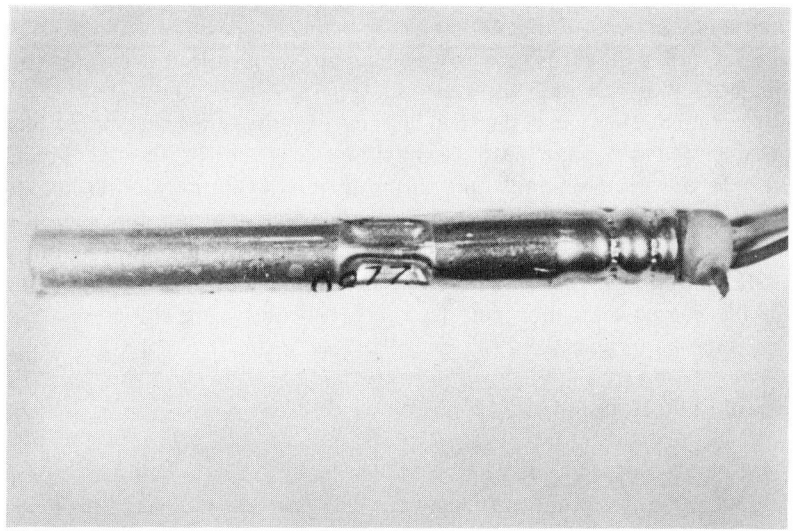

Figure 9.31   Water hammer effect on blasting caps.

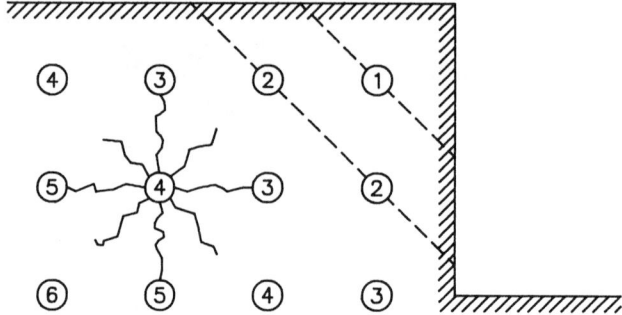

Figure 9.32   Disruptions caused by blastholes firing out of sequence.

column. Some of the explosives may burn, and some, depending on the degree of disruption, may not detonate. Improper drill hole spacings and burdens can also cause cracks to move in the wrong directions.

Execution of the Blasting Plan

The driller has a significant influence on whether or not blastholes will function properly, and whether they will release their full energy. Poor drilling procedures such as closely spaced holes can subject neighboring holes to extremely high-stress levels, and can cause problems such as dead pressing, precompression, and propagation and therefore, the pattern design, pattern execution, and the blasthole drilling are extremely important for proper results. The manner in which the blaster loads the holes also has an effect on performance. For example, insufficient stemming between decks within a hole invites propagation from deck to deck. Bridging of bulk explosives or cartridge hangup can also lead to loss of energy or unexploded explosive within the blasthole (Fig. 9.33).

When misfires occur in the field and one examines the cause of the misfire, it is sometimes found that the blasthole initiator was never tied into the circuit, and a missed hole was the problem. Care during hookup to insure that all initiators are properly tied in is extremely important. Another factor which has caused serious problems in blasting is the improper hookup of initiators. Many initiation systems used today require that the initiators be hooked together in a designated order to provide the proper detonation path from hole-to-hole. If the initiators are hooked together improperly holes will fire out of sequence.

Local Geology

The local geologic conditions can also have a detrimental effect on performance. Water in the blastholes changes the energy release for some explosives and can cause nonideal detonation. A loss of confinement, as a result of open seams or seams filled with sand or mud, can change the detonation characteristics of the

Figure 9.33 Holes venting as a result of cartridge hang-up near borehole collar.

explosives. These seams have the secondary effect of allowing the gas energy to be distributed in a totally different manner than intended around the borehole. This can lead to large boulders and little breakage. Seams can also be responsible for the traditional "cutoffs," or shifting of one portion of the powder column prematurely. The pattern design and initiator selection can intensify the negative effects of the local geology.

Geologic conditions. Geologic conditions are often very difficult to properly assess. No one has x-ray vision to look into the rock mass and assess geologic conditions. Even if this were possible, their affect on the blast performance cannot always be predicted accurately. There are times when geologic structure has a serious influence on breakage and other times when the same apparent structure seems to have little if any influence. The single most important geologic consideration is geologic structure. The jointing systems, dip and strike of bedding planes, mud or soft seams can have a serious influence on the blasting process both from a performance and safety standpoint.

Regional jointing patterns. The regional jointing pattern can influence both the breakage and overbreak in the blast. Figure 9.34 shows the directions of best and worst blasting along with the overbreak areas for a typical rock mass.

Regardless of the number of joint sets in a rock mass, one set of joints will be weakest or dominant and will control the breakage process.

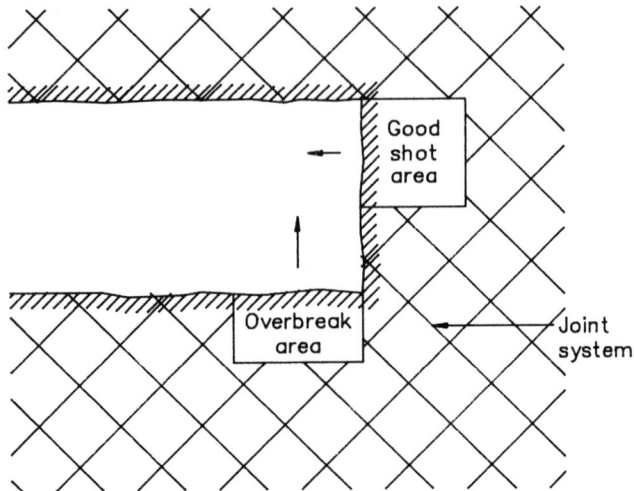

Figure 9.34 Effects of regional jointing patterns.

Dominant joints parallel the face.   If the dominant joints parallel the face fractures between, boreholes will prematurely link (Fig. 9.35). The premature linking will cause coarse or blocky burden fragmentation. Endbreak will be severe. Borehole spacing can be increased and fragmentation size will then decrease. If the intent of the blast is to produce rip-rap, a reduction of explosive load with close borehole spacing in the direction of the dominant joints will accomplish the task.

Joints perpendicular to face.   When the dominant jointing direction is perpendicular to the face as in Fig. 9.36, little endbreak will occur. However, backbreak will be significant. If large blastholes are used and many dominant joints occur between holes along a row, blocky breakage will occur between holes. The blocky breakage can be corrected by reducing spacing, but the backbreak may get worse as spacing is reduced. The use of smaller blastholes with a better distribution of explosive in the rock mass may be the best solution.

Joints at an angle with face.   When dominant joints are at an angle with the face, fragmentation is good and both endbreak and backbreak are normally within acceptable limits (Fig. 9.37).

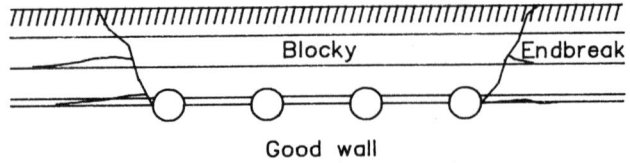

Figure 9.35 Dominant joints parallel to face.

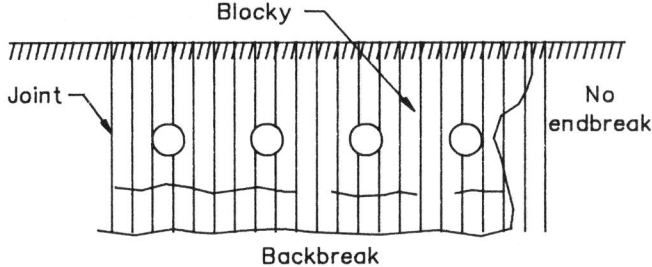

Figure 9.36   Dominant joints perpendicular to face.

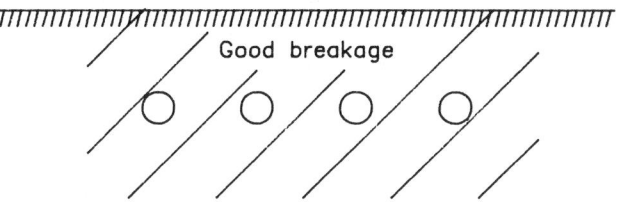

Figure 9.37   Dominant joints at an angle to face.

Joints at less than 30° angle to face.   Joints that form an acute angle with the face cause both breakage and wall stability problems. Figure 9.38 illustrates the results of this type of jointing. Burden breakage will be blocky and the back wall will be shattered, rough and broken.

Figure 9.38   Effects of joints at acute angles to face.

Blasting with the dip.   Figure 9.39 illustrates a through cut. On one side of the cut, the blast will be with the dip, while on the other side the blast will be against the dip.

When shooting with the dip, there will be more chance of backbreak. A smoother pit floor should result with less floor problems when bedding is closely spaced. The blasted rock should encounter less resistance and move further from the face. Burden distance can often be increased when shooting with the dip. When bedding layers are thick and massive, floor problems can result, and heavier bottom loads may be necessary to produce proper breakage.

When shooting against the dip, less backbreak will occur and the potential for rock overhanging the face will increase. The bottom of the shot will be more difficult to remove especially when beds are thick and massive and the floor can get rough. The muck pile will increase in height. The burden may need to be reduced to produce the desired floor conditions and fragmentation distributions.

Mud or soft seams.   Mud or soft seams cause more problems to blaster than any other geologic problem. They can occur in all types of rock. They are often unseen, yet if near a blasthole can cause severe violence and poor fragmentation.

Mud seams allow an almost instantaneous release of the explosive energy since they often move as a hydraulic fluid. Mud can be thrown far distances with flyrock traveling with the mud. Stemming across mud or soft seams is essential to obtain good blasting results.

Blasting in bedded rock.   Blasting parallel with the strike can produce results which are difficult to predict since many different rock layers can be inter-

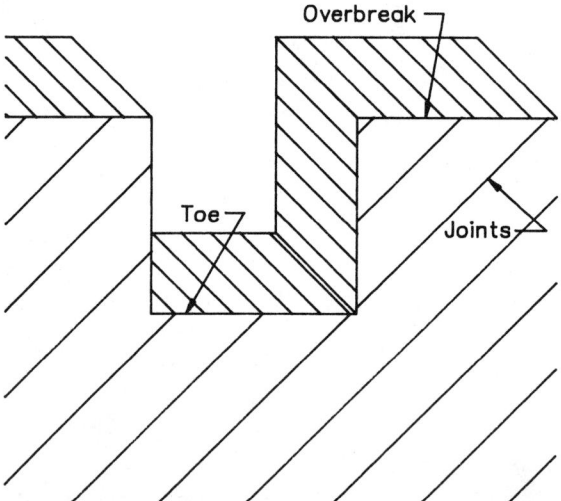

Figure 9.39   Considerations for dipping beds.

sected by a single pattern (Fig. 9.40). Since burdens and spacings are uniform in the pattern, layers will respond differently. Fragmentation will be different in each rock layer. Floor conditions and backbreak will differ in various sections of the blast.

In general, malfunctioning blastholes cause problems because there is too much resistance to rock movement at the time they fire, and therefore, poor breakage, airblast, flyrock, and heavy ground vibration result. This increased resistance on the blasthole at the time it fires is normally caused by either insufficient energy produced by the blast or energy released at the wrong time.

Problems with initiation timing.   The initiation timing is one of the most easily corrected causes of malfunctioning blastholes. Proper initiation time has many benefits, such as

1. increases in fragmentation with no additional explosive
2. reduction in vibration by as much as 75%
3. reduction in the instances of blowout and flyrock
4. added wall stability as a result of reductions in backbreak behind the shot
5. more control and more reproducible results in fragmentation and ground vibration.

Sufficient data exists to verify that timing does have this significant control on blasting results. The question is often asked as to why these results are only now

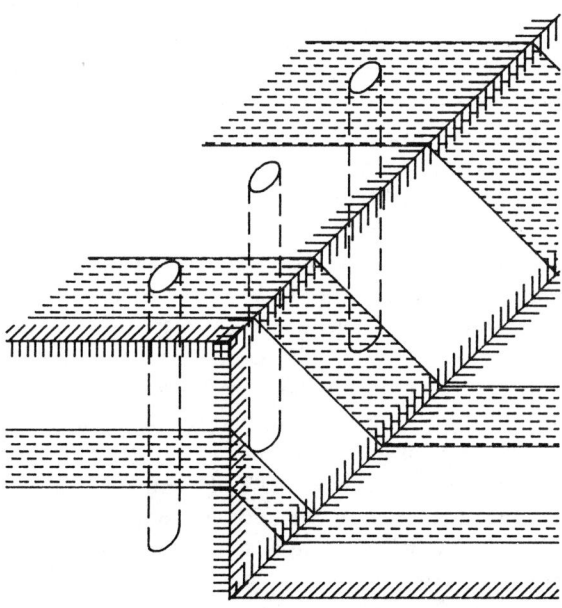

Figure 9.40   Blasting with dipping beds.

surfacing when millisecond delay caps have been made for over thirty years? The answer is because it was easy to design blasts if we assumed that caps fired at their nominal times.

The cause of variability in blasting results was blamed on ground conditions, changing geology, wet holes, or changes in types of explosives. Although these conditions could and were responsible for some of the variability, a great deal of the variability was caused by the inaccuracy of the initiators.

Correction of Problems

Malfunctioning blastholes are a common occurrence on all blasting jobs. Malfunctioning holes cause either no energy release, insufficient release, or energy released at an improper time. Malfunctioning holes also affect the performance of adjacent holes.

Malfunctioning holes do not always leave unexploded powder. However, when live explosives are found after a blast, we should not be satisfied that this is a random occurrence or an uncontrollable phenomenon. Instead we should investigate the cause. An unexploded explosive is only one symptom of a much greater problem with malfunctioning blastholes.

Effects on fragmentation.   To better understand the effect of malfunctioning blastholes on fragmentation, let us look at the effect of having a 40% overconfinement, which we will define as having 40% excessive burden on the hole at the time it fires. In another case, let us examine the effects of a 100% overconfinement, which would represent having twice the normal burden on the hole at the time it would fire. A 40% overconfinement could occur as a result of inaccuracies in drilling. The 100% overconfinement would simulate the effects of having a hole fire out of sequence, the second row firing before the holes in the front row. The burden from the first row has not had a chance to be displaced at the time the second row has fired and double the normal burden exists. We will use the "Breaker" software (1987) to analyze the effect of these conditions on fragmentation.

Figure 9.41 shows the dimensions of a typical blasting pattern. The 6.5 in. blastholes in limestone are used with a 15-ft burden. Figures 9.42 and 9.43 are numerical and graphical comparisons of the changes in fragmentation distribution resulting from a 15-ft (Entry 1) and 21-ft burden (Entry 2) which represents a 40% overconfined condition. The fragmentation size has increased with overconfinement and the size distribution has shifted toward the coarse as shown in Fig. 9.43. The average fragmentation size has increased from 14.9 to 19.6 in. The size distribution has shifted drastically, and in this case, 10% of the material is over 48 in. in size.

If a similar comparison in fragmentation distribution is made between a 15-ft burden (Entry 1) and a case where there is a 100% overconfinement (Entry 3), the numerical and graphical comparisons are shown in Figs. 9.44 and 9.45. The average size increased from 14.9 to 26.1 in. and also resulted in 27% of the material exceeding 48 in. in size.

```
Precision Blasting          BLAST FRAGMENTATION          Release:    2.01
Systems, Inc.               P R E D I C T I O N          Date: 03-30-1989
```

```
Mode:   INPUT/EDIT                          ACTIVE ENTRY #  1
```

```
Charge diameter [inches (decimal)] ....................    6.500
Burden [feet] .........................................   15.000
Spacing [feet] ........................................   19.000
Stemming [feet] .......................................   11.000
Subdrill [feet] .......................................    5.000
Bench height [feet] ...................................   60.000
Hole depth [feet] .....................................   65.000
Pattern factor (1 - for square, 2 - for staggered) ....    2.000
Number of rows ........................................    3.000
Specific gravity of explosive .........................    0.850
Rock strength (weak-fissured = 1, strong-massive = 10)..   3.000
Explosive charge per hole [lbs] .......................  648.000
Explosive volume strength (ANFO = 100) ................  100.000
```

Figure 9.41 Pattern dimensions.

```
Entry #  1 Calculation based on Volume Strength.
Entry #  2 Calculation based on Volume Strength.
                                    s c r e e n      s i z e
```

	3/16	3/8	3/4	1.5	3	6	12	24	48
% Passing									
Entry # 1	0.0	0.1	0.3	1.0	3.5	11.9	36.8	80.9	99.7
Entry # 2	0.1	0.2	0.7	1.8	4.7	12.1	29.1	60.2	91.6
% Fraction									
Entry # 1		0.1	0.2	0.7	2.5	8.5	24.8	44.1	18.9
Entry # 2		0.2	0.4	1.1	2.9	7.4	17.1	31.1	31.3

```
                                              Entry # 1      Entry # 2

            Average size [inches] ...........    14.996         19.628
            Fragmentation index .............     1.850          1.422
```

```
            Press:    B - Bar graph    L - Line graph    M - Menu
            To return to this screen from the Graph, press SPACE BAR.
```

Figure 9.42 Fragmentation distribution for 40% overconfinement.

Overconfinement of boreholes at the time of their detonation seriously affects fragmentation.

One of the most common reasons for blasthole malfunctioning, resulting in overconfinement, is the blasthole firing at the wrong time.

PROBLEMS

1. A blasting contractor will load 2-in. diameter dynamite cartridge ($SGe = 1.4$) into 2.5 in. holes in granite. The hillside that will be blasted will have bench heights that vary from 6 to 25 ft.

   a. Determine the spacing with instantaneous initiation along a row and a bench height of 6 ft.

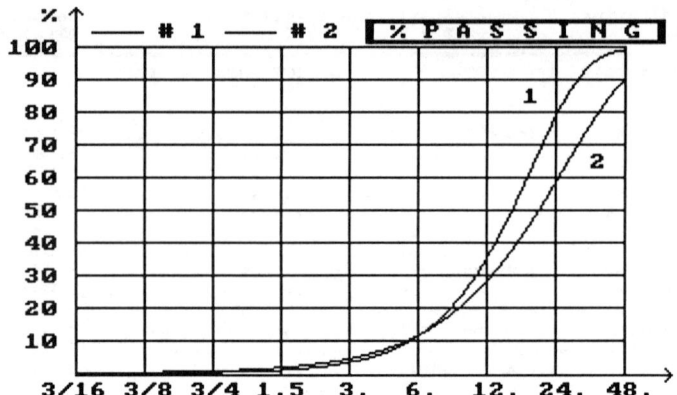

Figure 9.43 Fragmentation distribution for 40% overconfinement.

```
Entry # 1 Calculation based on Volume Strength.
Entry # 3 Calculation based on Volume Strength.
                                 s c r e e n      s i z e

          3/16      3/8      3/4     1.5     3      6      12      24      48

  % Passing

Entry # 1  0.0      0.1      0.3     1.0     3.5    11.9   36.8    80.9    99.7
Entry # 3  0.4      0.8      1.7     3.4     6.9    13.8   26.4    47.0    73.1

  % Fraction

Entry # 1            0.1      0.2     0.7     2.5    8.5    24.8    44.1    18.9
Entry # 3            0.8      0.9     1.7     3.5    6.9    12.6    28.6    26.1

                                            Entry # 1      Entry # 3

          Average size [inches] ...........   14.996        26.110
          Fragmentation index .............    1.050         1.048


      Press:    B - Bar graph     L - Line graph     M - Menu
      To return to this screen from the Graph, press SPACE BAR.
```

Figure 9.44 Fragmentation distribution for 100% overconfinement.

    b. Determine the spacing with instantaneous initiation along a row and a bench height of 25 ft.

    c. Determine the spacing for section 1a if delays are used along the row.

    d. Determine the spacing for section 1b if delays are used along the row.

2. You have been asked to design a sinking-cut in limestone quarry. A 6-inch blasthole will be used. The blastholes are dry, therefore, ANFO (*SGe* = 0.8) will be bulk loaded into the holes. The depth of cut will be 30 ft. Sixty-four blastholes (8 holes per row and 8 rows) will be used in the pattern. Design the shot, determine stemming distance, burden, spacing, subdrilling, and timing sequence.

3. 30 in. sewer lines must have 4 ft of cover. Rock is hard, massive sandstone, and houses

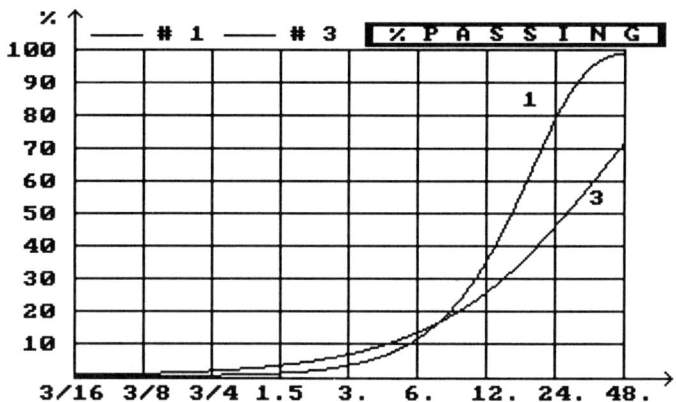

Figure 9.45 Fragmentation distribution for 100% overconfinement.

are within several hundred feet. We will drill 2.5 in. holes and use a 2-in. diameter Semigelatin Dynamite (S6e-13). Rock is wet. Determine drill pattern and explosive load.

4. Design the burden spacing subdrilling and stemming distance for a taconite ($SGr = 5.0$) mine which uses 15-in. diameter blastholes with 40-ft bench heights. A slurry explosive ($SGe = 1.1$) will be bulk loaded in the wet blastholes. Assume the blastholes will be fired instantaneously along a row.

5. Design a blasting pattern for a coal strip mine. The 10.625-in. blasthole will be bulk loaded with ANFO. The 50-ft deep overburden is composed of shale ($SGr = 2.5$). Homes are located at 1500 ft from the blasting site. Maintain PPV at 1 ips or less.

   a. Determine the maximum pounds of explosive per delay.
   b. Calculate the burden spacing and stemming distances.
   c. Determine the pattern dimensions needed to produce approximately 40,000 cu yd per blast with 4 rows of blastholes with a corner cut.
   d. Determine the timing, both hole-to-hole and row-to-row for an "average" condition.
   e. Calculate the total pounds of the explosive and total cubic yards blasted.
   f. Determine the expected ground vibration level at the homes.

REFERENCES

ANDREWS, A.B., "Design of Blasts, Emphasis on Blasting," Ensign Bickford Co., Simsburg, CT, Spring 1980, pp. 1, 4.

ASH, R.L. and KONYA, C.J., "Spacing: The Most Important Problem in Blasting," *Proceedings of Fifth Conference on Explosive and Blasting Technique*, February 1979.

ASH, R.L., KONYA, C.J., and ROLLINS, R.R., "Enhancement Effects from Simultaneously Fired Explosive Charges," *Transactions, SME/AIME*, 224:427–435, 1969.

"Blast Fragmentation Prediction (Breaker)," Computer Software Package, Precision Blasting Systems, Montville, OH, 1987.

CHIAPPETTA, R.F. and BORG, D.G., "Increased Productivity through Field Control and High-Speed Photography," First International Symposium on Rock Fragmentation by Blasting, Lulea, Sweden, 1983.

DRINKER, H.S., "Tunneling, Explosive Compounds, and Rock Drills," New York: John Wiley and Sons, 1882.

GILLETTE, H.P., "Handbook of Rock Excavation Methods and Cost," New York: McGraw-Hill Book Co., 1916.

GUSTAFFSON, R., "Swedish Blasting Technique," SPI, Gothenburg, Sweden, 1973, p. 323, available for consultation at Bureau of Mines Twin Cities Research Center, Minneapolis, MN.

KONYA, C.J., "Blasting Procedures at Woodville Lime and Chemical Company," *Proceedings of the Third Conference on Explosives and Blasting Technique*, Pittsburgh, February 1977.

KONYA, C.J., "The Effects of Joints and Bedding Planes on Rock Blasting," *Proceedings of the Second Conference on Drilling and Blasting*, International Society of Explosive Specialists, Phoenix, February 1973.

KONYA, C.J., "Problems with Malfunctioning Blastholes," *Proceedings of the Fourteenth Conference on Explosives and Blasting Techniques*, Society of Explosives Engineers, Montville, OH, February 1988.

KONYA, C.J., "Spacing of Explosives Charges," M.S. thesis, University of Missouri, Rolla, 1968.

KONYA, C.J. and FOLDESI, J., "A banyafal also reszenek jovesztesi problemai, nagy atmeroju nyujtott toltetek robbantasakor," Banyaszati es Kohaszati Lapok–Banyaszat 109(11): p. 728–732.

KONYA, C.J. and FOLDESI, J., "Kobanyaszati robbantasok Tervezese Nagytmeroju Nyujtott Toltetekkel," Epitoanyag, Budapest, Hungary, January 1977.

KONYA, C.J. and WALTER, E.J., Chap. 6, *Blast Monitoring, Surface Mining Environmental Monitoring and Reclamation Handbook*, Sendlein, L.V.A., et al, (ed.) New York: Elsevier Scientific Publishing Co., 1983, (in Press).

LANGEFORS, U. and KIHLSTROM, B.A., "A Modern Technique of Rock Blasting," John Wiley and Sons, Inc., New York: 1963, pp. 405.

LUNDBORG, N., PERSSON, P.A., LADEGAARD-PEDERSON, A., and HOLMBERG, R., "Keeping the Lid on Flyrock from Open Pit Blasting," *Eng. and Min. J.,* 176, no. 5, (May 1975), pp. 95–100.

PEARSE, G.E., "Rock Blasting—Some Aspects on the Theory and Practice," *Mine and Quarry Engineering*, 21(1):25–30, 1955.

PUGLIESE, J.M., "Designing Blast Patterns Using Empirical Formulas," U.S. Bureau of Mines, IC 8550, 1972, p. 33.

SCHAFFER, A., "A gyakorlati robbanto technika kezikonyve," Budapest: pallas Resvenytarsasag Nyomdaja, 1903.

SPEATH, G.L., "Formula for Proper Blasthole Spacing," *Engineering News Record,* 218(3):53, 1960.

VSETIN, Z., "Blasters Handbook," *Prague,* Czechoslovakia: Omnipol Prague, 1969.

10

Controlled Blasting

CONTROL TECHNIQUES

Blasting techniques have been developed to control overbreak at excavation limits. The operator must decide the ultimate purpose of the control technique before selection of the technique can be made. Some techniques are used to produce cosmetically appealing final walls with little or no concern for stability within the rock mass. Other techniques are used to provide stability by forming a fracture plane before any production blasting is conducted. This second technique may or may not be as cosmetically appealing, but from a stability standpoint, performs its function. Overbreak control methods can be broken down into three types: presplitting, trim (cushion) blasting, and line drilling.

Presplitting utilizes lightly loaded, closely spaced drill holes, fired before the production blast. The purpose of presplitting is to form a fracture plane across which the radial cracks from the production blast cannot travel. Secondarily, the fracture plane formed may be cosmetically appealing and allow the use of steeper slopes with less maintenance. Presplitting should be thought of as a protective measure to keep the final wall from being damaged by the production blasting (Fig. 10.1)

Trim blasting is a control technique which is used to cleanly shear a final wall after production blasting has taken place. The production blasting may have taken place many years earlier or could have taken place on an earlier delay within the same blast. Since the trim row of holes along a perimeter is the last to fire in a production blast, it does nothing to protect the stability of the final wall. Radial fractures from production blasting can go back into the final wall. Mud seams or other discontinuities can channel gasses from the production blast areas into the

Figure 10.1 Presplit blast on highway.

final wall. The sole purpose of a trim blast is to create a cosmetically appealing, stable perimeter. It offers no protection to the wall from the production blast.

Line drilling is an expensive technique, that under the proper geologic conditions, can be used to produce a cosmetically appealing final wall. It may, under proper circumstances, help protect the final contour from radial fractures by line drill holes acting as stress concentrators causing fracture to form between line drill holes during the production blasting cycle. If wall control was extremely important, one could not depend on line drilling to necessarily protect the final wall. Line drilling is more commonly used in conjunction with either presplitting or trim blasting rather than being used alone. Although the use of control blasting is more common for surface excavations, it has been successfully used underground, residual stress conditions permitting.

Principles of Operation

The explosive used for both presplitting and trim blasting is normally one which contains ammonium nitrate. Experience shows that high gas-producing explosives produce a better fracture and reduce the possibility of forming hairline cracks on borehole walls. The type of explosive used, however, is not critical. Most empirical formulas express the amount of explosives needed as the pounds of (any) explosive per foot of borehole. Common rules of thumb also indicate that the charge diameter be less than half the diameter of the hole. By using a small diameter charge in a larger diameter hole, the gas pressures drop quickly because of expansion into a larger volume. This procedure is called *decoupling*. This rapid drop in pressure has the effect of bringing explosive pressures into a narrow range of values for most types of common explosives used. In effect what occurs is that under the proper decoupling, different explosives produce stresses in the rock which are approximately within 10% of one another in a presplitting or trim blasting application. An example of the stresses produced 12 in. from the blasthole is given in Fig. 10.2. The

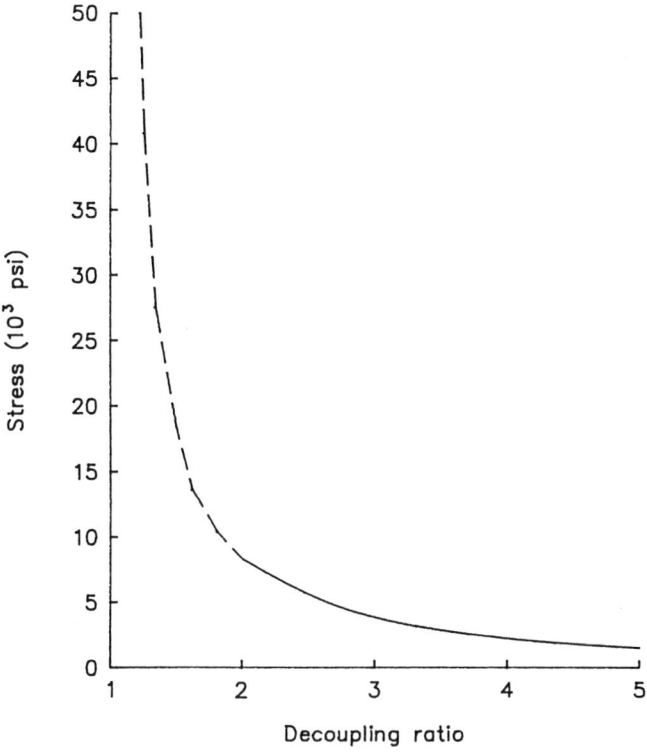

Figure 10.2   Stress levels from decoupled shots 12 In. from blasthole walls.

decoupling ratio is defined as the diameter of borehole divided by the diameter of charge.

Past explanations of presplitting indicate that it was caused entirely by the reflection of stress waves as shown in Fig. 10.3. Later research proved that the

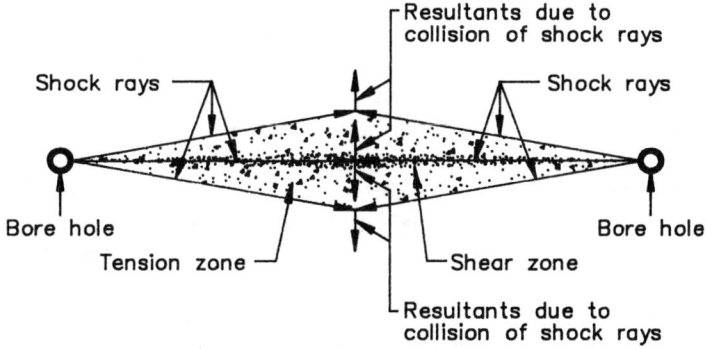

Figure 10.3   Old concepts of stress wave breakage.

magnitude of the resultant stress is insufficient to cause the splitting action to occur in real blasting situations. If one had to rely only on the stress waves to cause presplitting, spacings would have to be reduced to 1/5 those which are commonly used in the field. If blastholes within a presplit row were not fired truly instantaneously according to Fig. 10.3, the splitting action could not possibly result, since stress wave collision would not occur between holes. This is contrary to fact, since blasters commonly delay each hole in a presplit shot and still produce good wall conditions. Figure 10.4 shows a presplit forming from radial crack growth, not stress wave collision. The high speed photographs were taken 2 millionths of a second apart. The model material was plexiglas and the explosive charge was PETN.

Figure 10.5(a) is a photograph of a Plexiglas model in which three blastholes were fired simultaneously. Figure 10.5(b), on the other hand, is a photograph of a model where blastholes were fired on what would be equivalent to a 25-ms delay in full scale blasts. There is no significant difference in breakage between holes. Tests have shown that stress wave interactions are not responsible for the splitting in full scale blasts.

This point is significant because, if you would believe in the stress wave breakage concept as being the prime mechanism for presplit formation, then all blastholes along the perimeter of an excavation would by necessity be fired instantaneously. Since the presplit is normally the closest to residences and also the most heavily confined holes in the entire blast, high-vibration levels would be produced per pound of presplit explosive used. Levels could be as much as five times higher than those in production blasting. Many holes fired instantaneously would cause excessively high-ground vibrations. The fact that holes can be delayed is important because it allows you flexibility to fire each hole on a separate delay if necessary.

Presplitting is not a new blasting technique. It became a recognized blasting technique for wall control when it was used in the mid-1950s on the Niagara power project. Its use was reported as early as the 1940s on a sporadic basis.

Presplitting was used as a rock fracturing technique before explosives were used for blasting. The pyramids of ancient Egypt were built by craftsmen that used a nonexplosive method of presplitting. The technique was employed by pounding wooden wedges into holes drilled into the rock. The wooden wedges were soaked with water and the forces generated by the wood expansion caused fractures to. occur between holes. The blocks could then be removed.

In northern climates, man found that he could use the forces generated by freezing water to cause rock to fracture. Holes were drilled into a rock mass and filled with water. Cracks developed between holes as the water froze during the winter. In the spring, the blocks could be moved. Both the wooden wedges and the freezing water exerted static pressure on the rock mass similar to what occurs from the explosive gas pressure.

Empirical formulas used in presplitting normally do not take into consideration strength characteristics of the rock mass. Although this may seem unusual, it must be remembered that tensile strength ranges from a few hundred to no more

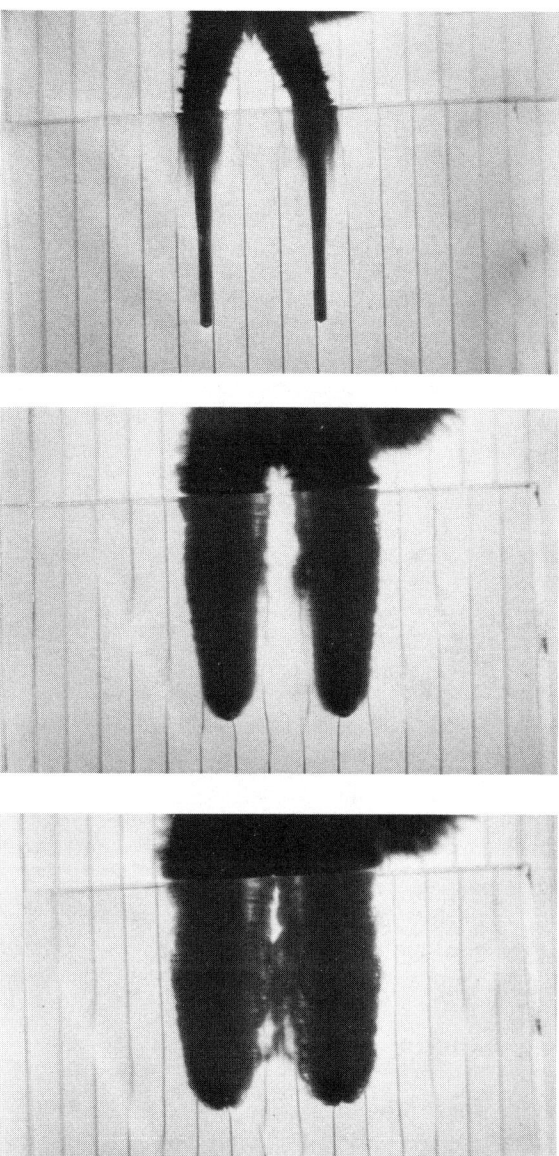

Figure 10.4 Presplit fracture formation in plexiglas models. (a) Shock wave collision in collar (no crack formation); (b) Radial cracks begin to accelerate between holes; (c) Radial cracks link to form presplit.

than a few thousand psi in rock. Crushing strength, on the other hand, is normally rated in tens of thousands of psi. If the explosive pressure within the blasthole is such that it is below the crushing strength and above the tensile strength, fractures

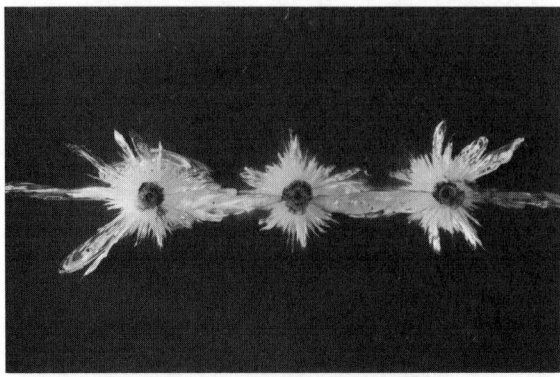

Figure 10.5   Three-hole presplit. (a) Instantaneous firing; (b) delay firing.

will occur without damaging the rock mass around the borehole. In most presplitting and trim blasting applications, pressures generated are between 8 to 15 thousand psi and vastly exceed the tensile strength of any rock. Therefore, the tensile strength would not be a major consideration.

Presplitting with Propellants

The method by which presplit fractures are formed is still a matter of debate by some practitioners. In order to take the mechanism of presplitting out of the realm of theory and small scale laboratory tests, we will describe the full scale presplit blasts conducted with a propellant. The propellant, Pyrodex, deflagrates and does not detonate. Blasts were conducted in granite quarries where previous presplit blasting had been done using dynamites and other high explosives. These blasts showed successful presplits could be made on widely spaced holes with loads that are equivalent to that which would be used with dynamite. The significance of these tests are that they conclusively prove that in field blasting applications good presplits can be formed with explosives that produce no stress wave whatsoever.

When Pyrodex reactions are compared to those of high explosives used for presplitting, the unique beneficial characteristics are evident.

High explosives, upon detonation, react completely before venting of high-pressure gases can occur. If the holes are overloaded, damage results and multiple fractures form rather than only one linking the holes. The wall rock is irreparably damaged. Pyrodex has a low velocity upon loss of confinement. It slows down immediately when the first presplit fracture is formed regardless of the amount of overload. Therefore, one can load a borehole with Pyrodex and not be overly concerned if the hole is somewhat overloaded.

Good stemming is important to the performance of Pyrodex since its reaction rate slows down when the confinement is relieved.

Test results in Georgia granite.   Full scale presplit blasts were conducted in Georgia granite with Pyrodex and compared with previous presplit blasts conducted with dynamite. The tests were conducted in a granite quarry near Atlanta, Georgia. The granite was massive and homogeneous and used to produce dimension stone. The reason this site was selected was because we had experience and data in shooting this stone with dynamite charges. Blast holes were drilled approximately 10 ft deep and on 3-ft centers. It was decided after the holes were drilled to load only every other blast hole. Each loaded hole contained 3 lb of Pyrodex and was initiated by a single instantaneous electric blasting cap. The loaded holes were spaced on 6-ft centers with an unloaded hole between each pair of loaded holes (Fig. 10.6).

The test results showed that Pyrodex did cause presplitting even though holes were placed on 6-ft centers with an empty hole located 3 ft from the loaded holes (Fig. 10.7). Figure 10.8 shows that unburned powder is blown from the split and reacting in air indicating that the crack occurred without the full utilization of all the Pyrodex and that the splitting action could have been accomplished using a smaller powder load. The Pyrodex indeed slowed to a very low reaction rate once the split formed and the explosive lost confinement. Typical dynamite shots in the same material were loaded using 3.25 in. blastholes drilled on 2-ft centers approximately 10 ft deep. Into these holes were placed distributed dynamite charges with a total weight of approximately 0.75 lb per hole. Three feet of stemming was used at the collar of each hole. The holes were fired air cushioned with stemming only in the top 3 ft of the hole. The results with the ammonia gelatin dynamite charges are shown in Fig. 10.9. The holes loaded with dynamite were initiated using detonating cord. The comparison of the results are in Table 10.1. For a 6-ft section of face approximately 2.25 lb of dynamite were used. This would result in approximately 0.04 lb of dynamite per sq ft of new face developed by the presplit crack.

At boarder spacings, typical dynamite shots on 44 in. centers would require charges of about 0.27 lb per ft or about 2.44 lb per hole. The amount of explosive per square foot of face would be 0.07 lb per ft.[2]

Three pounds of Pyrodex was used in the loaded holes spaced on 6-ft centers. The explosive load per square foot of face produced was 0.05 lb or well within the range of charge weight expected to accomplish the same task with dynamite.

Figure 10.6   Presplit pattern in Georgia granite. Single row on 6 ft. centers.

Mechanism of fracture formation.   Under field conditions Pyrodex reacts at 1600 ft per sec, which is subsonic in the rock. This is quite different from the dynamite charges since the dynamite has a detonation velocity of approximately 13,000 ft per sec and is accompanied by a shock wave which formed stress waves in the granite. Presplit fractures were formed in tests in granite and the amount of explosives per square foot of fracture face developed was the same with dynamite and Pyrodex. Therefore one pound of Pyrodex caused similar results as one pound of dynamite. Pyrodex, however, produced no stress wave. As a result of these full scale tests one can only conclude that under full scale field conditions the stress wave is of little benefit in the creation of presplit fractures.

The results of these controlled field tests should produce a better understanding of the role of the stress wave and that of sustained gas pressure in producing fractures in rock.

Figure 10.7   Presplit fracture formed with propellant.

Air Cushion Blasting for Presplitting

A technique which was previously discussed under air cushion blasting for boulders is also used in wall control. This technique utilizes explosives placed only at the bottom of the blasthole with the remainder of the hole left empty except for

TABLE 10.1   Comparison of Presplit Results from Dynamite and Pyrodex

Hole Diameter (in.)	Hole Spacing (in.)	Product	Charge Hole (lbs.)	Explosive per ft.^{2} of face (lb/ft.^{2})	Drill feet per ft.^{2} of face
2.75	24	Dynamite	0.75	0.04	0.5 (presplit)
2.75	44	Dynamite	2.44	0.07	0.27 (presplit)
3.25	72	Pyrodex	3.00	0.05	0.33* (presplit)
1.63	8	Pyrodex	0.56	0.07	1.5 (dim. stone)

*includes unloaded hole

Figure 10.8   Ejection of burning explosive.

Figure 10.9   Presplit fractures formed with dynamite.

stemming at the collar. This technique dates to the use of black powder in dimension stone. The stemming at the collar is held in place by a balloon or other materials which could lock into the borehole walls below the stemming zone. The stemming is then placed above these obstructions. When the explosive detonates in the bottom of the hole the gas pressures quickly equalize uniformly stressing the borehole walls. The advantage of using this technique is that larger diameter boreholes can be spaced further apart with a less expensive explosive such as ammonium nitrate used as the presplitting charge. By using air cushion blasting where applicable, the cost of presplitting is reduced. The technique, however, is not universally applicable and may in some circumstances, not perform as well as closely spaced lightly loaded presplit holes. One reason the technique is less reliable is because the broader spacings between holes allow local geology to play a more significant role in hindering the proper formation of the presplit fracture.

Effects of Local Geologic Conditions

Control techniques such as presplitting, trim blasting and line drilling work best in massive rock. You can see the *half casts* or half of each borehole on the final wall. In massive rock, almost 100% of the holes produce half cast. Some operators try to assess the success or failure of presplit or trim blasts by what is called a half cast factor. Half-cast factors are the percentage of the total half casts which are visible after the rock has been excavated. If only 40% of the drill holes remain visible on the final wall as half casts, then the half cast factor would be 40%. This technique could have some merit when blasting in solid homogeneous massive material. However, half casts may totally disappear in geologically complicated rock. One cannot assume, however, that the lack of half casts indicates a poor blasting job. As an example, in geologically complicated material a clean crack does not form, however, there is a broken shattered zone formed along the perimeter. That zone serves its function of protecting the final wall from the effects of radial cracks emanating from the production blast. Half cast factors, therefore, only have validity if the rock type in which the half casts are being counted are considered in the evaluation.

When rock has numerous joints between blastholes and those joints intersect the face at less than a 15° angle, it may be impossible to form a good looking face with control blasting techniques. Dominant joints must intersect the face at greater than a 30° angle to produce reasonably straight faces. Anything less will cause fractures to intersect the jointing planes having large pieces of material fall out from the face during the excavation process.

In a weak material, the skill of the excavator operator is critical. Some machines can exert considerable thrust, whereby they can dig into an unblasted wall severely damaging the final contour. Other geologic factors which effect the outcome of control blasting techniques are soft seams or mud seams. If the bench is intersected by numerous mud seams it is difficult to produce good results.

Presplitting

In order to prepare a presplit blasting plan which is within normally accepted limits for our test shot, we could use the following equations.

The powder load per foot of hole which will not damage the wall but will produce sufficient pressure to cause the splitting action to occur, is given by equation 10.1:

$$dec = Dh^2/28 \qquad (10.1)$$

where

dec = explosive load per foot, lb/ft
Dh = diameter of empty holes, in.

If this powder load is used, the spacing between holes in a presplit blast can be determined by

$$S = 10Dh \qquad (10.2)$$

where

S = spacing, in.
Dh = diameter of the empty hole, in.

The constant 10 in the above formula is conservative. It is used to make sure that the presplit distance is not excessive and that the presplit will occur. Field experience indicates that often this value can be increased to 12 and sometimes 14.

In most presplitting applications there is no drilling below grade. However, a concentrated charge, which is equivalent to approximately 2 or 3 times *dec,* is placed in the bottom foot of the blasthole. The blasthole should be fired either instantaneously or on a short delay between each hole. Although some contractors have reported satisfactory results with 50 ms delays, it is not recommended to delay greater than 25 ms between holes.

A presplit shot is meant to cause a fracture to occur and travel to the surface of the ground. If this occurs, no amount of stemming placed in the hole will hold and it will be ejected. Therefore, drill cuttings can be used safely as stemming since its function is to momentarily confine the gasses and to cut down on some of the noise. Normally, holes are stemmed in the top 2 to 3 ft depending on their diameter. The larger the hole diameter, in general, the more stemming is used.

The question as to whether or not stemming should be used between charges in the hole is one where there are differing opinions. The author recommends the following. If the rock mass to be blasted is seamy in nature and has many partings and small seams, it might be wise to place stemming between charges. On the other hand, if the rock mass is competent, although it may be bedded, stemming between charges is not necessary, especially in materials that have a very low crushing strength such as weak shales. Leaving an air cushion around charges can be beneficial. By not stemming around charges, a greater empty volume is available for the explosive gas expansion, thereby dropping the gas pressure more quickly. The pressure per square inch is lower; however, more square inches of the hole are being stressed and therefore, good fracture results. When stemming is used between charges in weak rock, the walls can be pock-marked where the high pressure gases were released at the charge location.

Explosives for presplitting come in many types. There are polyethylene coils which are snaked down the hole in diameters less than an inch. These polyethylene tubes contain slurry explosives. Other types of charges are slender dynamite cartridges which couple together as they are put down the hole to form a continuous charge. Other methods of placing charges consist of taping either full or fractions of dynamite cartridges to a detonating cord and lowering that assembly into the blasthole. The choice of which charges to use depends on the operator and what is available in his area. What is important is that the charges be less than half the diameter of the blasthole, and preferably not touching the blasthole walls.

An example of the use of these formulas for determining the loads and distances for a presplit shot is given in Ex. 10.1.

Example 10.1

A presplit blasting plan is to be developed for a highway project. The plan calls for 3-in. blastholes. What explosive load and drillhole spacing should be used for the test blast? Holes will be fired near instantaneously with detonating cord.

Determine powder load

$$dec = Dh^2/28$$
$$dec = 3^2/28 = 0.32 \text{ lb/ft}$$

Determine spacing

$$S = 10Dh = 10 \times 3 = 30 \text{ in.}$$

Bottom load (*deb*)

$$deb = 3 \, dec$$

$$deb = 3 \times 0.32 = 0.96 \text{ lb} \tag{10.3}$$

Some operators prefer to load the production holes nearest the presplit line lighter than they would load the remainder of the production holes. The first row of buffer holes, as they are commonly called are often closer spaced, with smaller burdens and lighter loads so that less pressure will be placed on the final wall.

Trim (cushion) Blasting

Trim blasts are fired after the production round has been fired. They are designed in a similar manner to presplit blasts. The powder load per foot of hole is determined by Eq. 10.1 just as we did in presplitting. The spacing is normally larger than you would use in presplitting because there is relief toward which the holes can break. As an example, the following equation could be used to determine the approximate spacing for a trim blast.

$$S = 16Dh \tag{10.4}$$

where

$$S = \text{spacing, in.}$$
$$Dh = \text{diameter of the empty hole, in.}$$

Confinement conditions are different when trim blasting than when presplitting. During presplitting, the production round has not yet fired and for all practical purposes, the burden is infinite. In trim blasting, however, burdens are normally within reasonable distances since the production round has been fired. The burden must be considered in the design of a trim blast. The blast is designed so that the burden is greater than the spacing to ensure that the fractures will properly link between holes rather than prematurely moving toward the burden. The following equation is commonly used:

$$B = 1.3S \tag{10.5}$$

where

$$B = \text{burden, in.}$$
$$S = \text{spacing, in.}$$

Stemming considerations both at the collar of the blasthole and also around the charges for trim blasting would be the same as those for presplitting. In the trim blast application, subdrilling is not normally necessary. However, concentrated bottom loads to cause the cracks to go to the grade line are normally used. These bottom loads can be determined in the same fashion as was described under presplitting. Example 10.2 shows how a trim blast can be designed.

Example 10.2

A contractor will use a 2.75 in. blasthole for a trim blast. Determine the blasting plan. Determine powder load (Eq. 10.1)

$$dec = Dh^2/28 = (2.75)^2/28 = 0.27 \text{ lb/ft}$$

Determine spacing (Eq. 10.4)

$$S = 16Dh = 16 \times 2.75 = 44 \text{ in.}$$

Determine bottom load $= 3 \times dec = 3 \times 0.27 = 0.81$ lb
Check minimum burden (Eq. 10.5)

$$B = 1.3S = 1.3 \times 44 = 57 \text{ in.}$$

Trim blasting with detonating cord. In some applications where trim holes must be drilled at very close spacings, normal charges are too large and cause overbreak around the holes. The use of closely spaced holes, on 12- to 24-in. centers, may be necessary in some geologic formations and for concrete removal on some excavations. In some cases it is also necessary to drill larger holes than normally would be used, however, the spacings are short. Additional airspace around the charges is not normally detrimental to the formation of the split. If, however, we use the equations based on the hole diameter to calculate the loads, the charges would be too large for the spacings. On these close spacings we then can use the formula 10.6 to determine the amount of explosive which would be necessary for a fixed close spacing. It is often convenient to use detonating cord to provide this small distributed load.

$$dec = 7000 \ (S/85)^2 \tag{10.6}$$

where

$$dec = \text{loading density in grains/ft}$$
$$S = \text{spacing in in.}$$

Example 10.3

Two-inch diameter blastholes will be drilled on 18-inch centers and 20 feet deep. Determine the grain load of detonating cord needed to shear the rock web on the trim blast.

$$dec = 7000 \ (S/85)^2$$
$$dec = 7000 \ (18/85)^2$$
$$dec = 314 \text{ grains}$$

Line Drilling

Line drilling is a fracture technique which utilizes empty drill holes spaced within 2 to 4 diameters from one another. These unloaded closely spaced drill holes under proper geologic conditions can act as stress concentrators or crack guides to cause cracks to form between them. Unloaded line drill holes are sometimes used in tight corners to guide cracks from a presplit into a specific angle. Line drilling is also employed between presplit or trim blastholes to help guide the cracks between loaded holes. In geologically complicated material, however, line drilling may not function as desired, since fractures tend to concentrate at naturally occurring weakness planes rather than at the man-made weakness plane created by the line drilled holes.

Assessment of Results

The above formulas are guidelines which are used to calculate the powder loads and spacing for controlled blasting techniques. After test shots are conducted, the operator can evaluate the results and determine whether changes are needed in the blasting plan.

If the rock is massive with few geologic discontinuities, too great or too little spacing can be assessed by looking at the fracture plane formed. Figure 10.10 indicates the results that would be obtained if blastholes are spaced too close for the powder load used. Numerous fractures link in the plane between holes and when the blast is excavated the material between holes will fall out leaving half casts protruding from the final wall. If the powder load is too great and holes are overloaded, crushing of the borehole wall will result. If spacings are too far, a face that is generally rough in appearance will result (Fig. 10.11).

Crushed zone

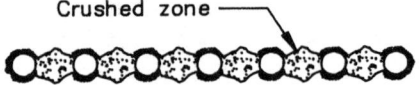

Final wall

Figure 10.10   Presplit effects with close spacing.

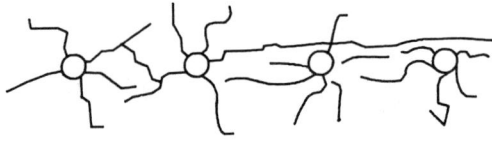

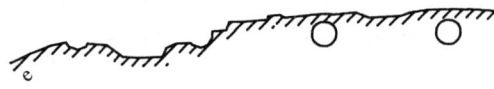

Final wall

Figure 10.11   Presplit effects with large spacing.

If rock is not massive but contains numerous near vertical tight joints inter-secting the face, the results will be different. If the tight joints intersect a line between holes at a 90° angle, the break line should be relatively straight between the face (Fig. 10.12). If the dominant joints intersect the face at an acute angle, breakage as indicated in Fig. 10.13 will result. This type of breakage, which leaves the half-cast protruding from the final face, would seem to indicate that boreholes are spaced too close and overloaded. Boreholes may be properly spaced, but the acute angles formed between the dominant joints and the face cause a different fracture pattern with two or more fractures linking between holes (Fig. 10.14).

If joints approach the face at less than a 15° angle, the face produced by the control technique may show no half casts whatsoever and may appear to be rough and torn (Fig. 10.15). Little can be done in this situation. Figure 10.16 shows a presplit face with joints at an acute angle. Few half-casts remain. If joints are weak and at an acute angle, backbreak will result. If in addition these joints are dipping, a ragged unstable condition can result. Figure 10.17 shows the presplit results with these geologic problems. Although not cosmetically pleasing, the face in Figure

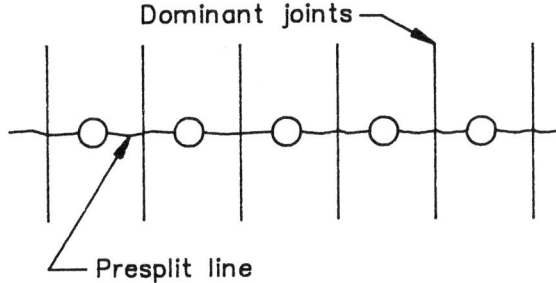

Figure 10.12   Presplit with joints at 90° angle.

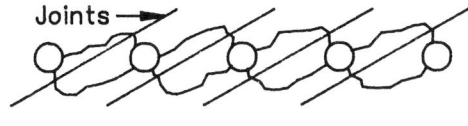

Final wall

Figure 10.13   Presplit with joints at acute angles.

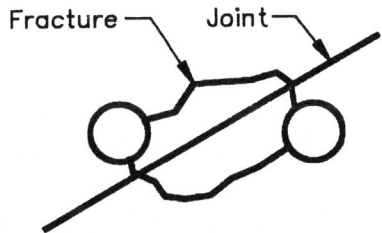

Figure 10.14   Breakage diagram for pre-splitting in jointed rock (after Worsey).

10.16 should be stable. This type of geologic structure may promote raveling of the face; however, mass movement because of instability should not occur as a result of blasting.

Causes of Overbreak

Two general types of overbreak occur from a production blast. *Backbreak*, the breakage behind the last row of holes and *endbreak*, the breakage off the end of the shot.

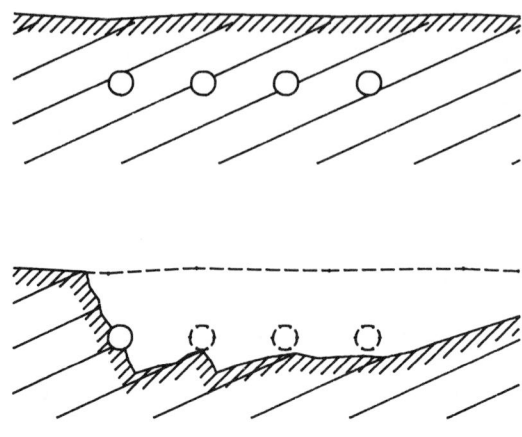

Final wall

Figure 10.15   Rough wall which results from presplitting with joints at acute angles.

Backbreak.   There are many causes of backbreak. It can be due to excessive burden on the holes thereby causing the explosive to break and crack radially further behind the last row of holes (Figure 10.18). Improper delay timing from row-to-row can cause backbreak when the timing is too short because there is

Figure 10.16   Presplit with joints at an acute angle.

Figure 10.17   Acute angled dipping joints cause ragged faces and unstable walls.

excessive confinement on the last rows in the shot. Benches that are excessively stiff (L/B < 2) cause more uplift and backbreak near the collar of the hole (Fig. 10.19). Long stemming depths on short benches also promotes backbreak. If blastholes are short, with low L/B ratios because of excessive burden, the obvious solution to the problem would be to change to smaller holes thereby reducing the burden and increasing the stiffness ratio. This procedure, however, cannot be followed in all operations. Therefore, other techniques must be used to cleanly shear holes at their collars.

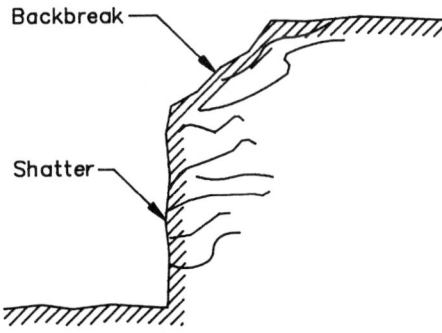

Figure 10.18   Backbreak caused by excessive burden.

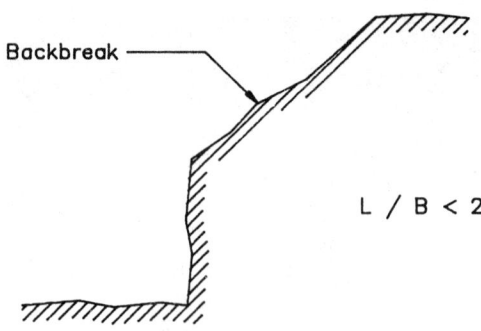

Figure 10.19 Backbreak caused by excessive stiffness.

Satellite holes can be used between the production holes whereby the cap rock in the area of the stemming zone can be lightly loaded and fired on a later delay. Operators often drill satellite holes (Fig. 10.20) which helps reduce problems with cap rock and reduce overbreak. When satellite charges are used within the stemming zone as indicated in Fig. 10.20, those charges should be fired on a short delay after the main charge shoots. One would not want to prematurely unconfine the main charge in the blasthole by having the satellite charge fire first and blow out the stemming.

Another technique similar to using satellite charges is to continue the main charge into the stemming zone. The main charge, however, is reduced in diameter. This small diameter charge in a much larger hole produces sufficient pressure to cause some cracking similar to presplitting in the collar area (Fig. 10.21).

Endbreak.   Endbreak off the end of a shot usually results from one of two reasons (Fig. 10.22). The local geologic structure promotes extension of cracks off the end of the shot. This can be helped by shortening the spacing distance on the end

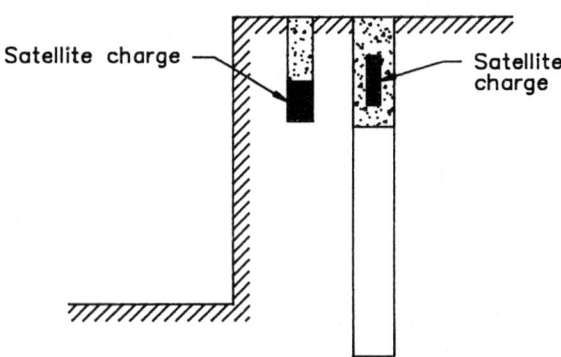

Figure 10.20 Satellite charges in collar.

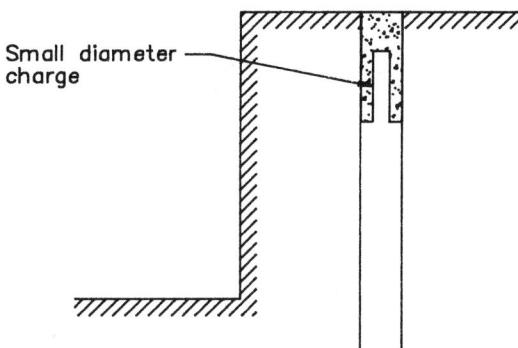

Small diameter charge

Figure 10.21 Charge extended into stemming.

between the nearest production holes thereby causing the hole to function and respond in a different fashion.

Endbreak can also be caused by having improper timing on the perimeter holes. If the timing is too fast, blastholes will sense a much larger than normal burden thereby either rifling and causing uplift, or by cracking back into the formation. The problem of timing can be corrected in the same manner as that described for backbreak, whereby longer delay times, such as those which were previously discussed in Chap. 9, can be used on the end holes, allowing time for the center portion of the blast to begin to move out, thereby producing additional relief before the end holes fire.

Flyrock control.   In general, flyrock results from one of two places in the shot. It either comes from the face or it comes from the top. If flyrock is originating from the face and flying far distances, it could be an indication that too little burden is used or that mud seams or other geologic discontinuities are prevalent. Most flyrock, however, is not produced from the face. It is produced from the top of the shot. It results from geysering or vertical cratering of holes. Geysering of blastholes normally results from overconfinement of holes at the time they fire, because of poor initiation time selection. Vertical cratering can result for similar reasons, it can also occur because of careless loading where explosive columns are either brought

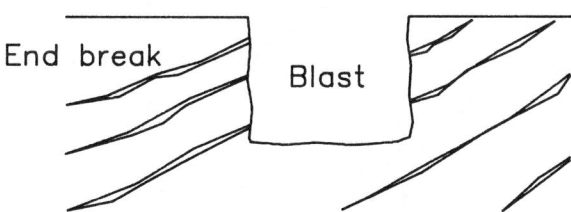

End break   Blast

Figure 10.22 Endbreak (plan view).

up too high in the hole or if a powder cartridge becomes lodged in the stemming zone during loading and insufficient stemming is used. Care in loading would solve both problems before they occur. The timing problem is similar to what has been discussed for backbreak and endbreak. Increasing the time between rows of holes within the blast should solve the problem.

There are times when to produce proper breakage, one must deliberately load higher and heavier in the hole than would normally be required. These situations result when very low L/B ratios occur in massive rock. In these cases, when small diameter blastholes are deliberately slightly overloaded to promote top breakage, one can use 3 to 4 ft deep soil covering over the shot to act as a blasting mat to restrain potential flyrock. Blasting mats made of woven wire or wire and rubber tires can also be placed on top of the shot both with and without earth mats to contain the flyrock.

PROBLEMS

1. The perimeter along a highway must be presplit for stability. The depth of the cut will be 40 ft. A track drill with 3.0 in. bit will be used to drill the blastholes.
   a. Determine the loading density of the explosive.
   b. Determine the total explosive load per hole.
   c. Determine the spacing.
2. You are responsible to trim blast a 40-ft limestone face. The specific gravity of the limestone is 2.6. The drilling holes will be 4.0 in. in diameter.
   a. Determine the loading density.
   b. Determine the spacing of blastholes.
   c. Determine the necessary burden.
3. You have been asked to design a trim blast with detonating cord as the explosive with 3.5 in. blastholes drilled on 24 in. centers. Determine the grain load of detonating cord needed to do the job properly.

REFERENCES

E.I. DUPONT DE NEMOURS & Co., Inc. "Blaster's Handbook," 16th ed., Wilmington, DE, 1978, p. 494.

E.I. DUPONT DE NEMOURS & Co., Inc. "Four Major Methods of Controlled Blasting," Wilmington, DE, 1964, p. 16.

KONYA, C.J., "High-Speed Photographic Analysis of the Mechanics of Pre-Split Blasting," *Proceedings of Sprengtechnick International,* Linz, Austria (in German), 1973.

KONYA, C.J., "Presplit Blasting: Theory and Practice," Preprint, AIME, Las Vegas, 1980.

KONYA, C.J., BARRETT, D., and SMITH, E., Jr., "Presplitting Granite Using Pyrodex, A

Propellant," *Proceedings of the Twelfth Conference on Explosives and Blasting Techniques,* Society of Explosives Engineers, Montville, OH, 1986.

WORSEY, P.N., "The Effects of Discontinuity Orientation on the Success of Pre-Split Blasting," *Proceedings of the Tenth Conference on Explosives and Blasting Technique,* 1984, pp. 197–217.

11

Standards and Procedures for Environmental Control

Setting Up the Seismograph

Where to place the seismogragph is the first item of business. A complaint or an area of concern that needs to be checked frequently will decide this. If there is no such problem, then place the seismograph at the nearest structure that is not owned by or connected with the operation. The seismograph distance should always be less than or, at most, equal to the distance to the structure of concern.

Seismograph measurements are made for the purpose of reducing hazard and the potential for damage. In dealing with residents or persons in the vibration affected area, be factual and direct. Emphasize that the purpose of the seismograph measurement is to protect them and their property from vibration damage, and that standards have been developed by the Federal Government and other scientific investigators to do this.

Placement of the sensor is important. Place it on solid ground. Do not place it on

grass

loose earth

any soft material

an isolated slab of stone or concrete

inside a structure except on a basement floor

concrete or driveway connected to a blast area.

Biased vibration readings that are not representative of the true ground vibration probably will occur if these precautions are not observed.

Level the sensor using the level bubble on the unit. Some units do not have level bubbles in which case it can be leveled by eye.

The sensor must be solidly planted so as to maintain contact with the earth. It should vibrate as part of the earth, not as an independent mass. When ground motion is large, special precautions may be called for such as, cover the sensor with a sand bag, spike it down or dig a hole and cover it with earth. If the sensor is merely setting on the ground, large motion may decouple it from the earth and the vibration record will not represent the true ground motion. The ground displacement is usually only a few thousandths of an inch so the decoupling motion will not be apparent to the seismograph operator. The decoupled motion will be obvious on the seismogram.

Sound measurement—most sound measurement is made with a hand held microphone. Hold the mike at arm's length to the side away from you to avoid reflection of the sound wave from your body. If available, set the microphone on a tripod stand. Also avoid setting the microphone in front of a building or a wall to avoid reflection of the sound. Reflection of the sound could double the sound-level pressure.

Modifying the Scaled Distance

An effective blasting program based on the latest technology and sound engineering may indicate using a scaled distance value outside that allowed by regulatory prescriptions. It is necessary then to justify the use of the modified scaled distances as safe for structures affected by the vibration. Seismograph measurement of the blast vibration will have to be made.

Methods Averages

A series of test shots or production shots can be used. A minimum of five will serve as a start. Compute the scaled distance value for each blast. Seismograph recordings must show these blasts to be safe. A safe blast is one that is within regulatory particle velocity specifications or scaled distance specifications. Using the scaled distance values for the blast series, calculate the average scaled distance. This average scaled distance, when approved by the regulatory agency, may then be used as the operational control value on succeeding blasting and also to serve as a basis for computing safe charges and safe distances.

Example of calculation

SHOT	D	W	$W^{0.5}$	Ds
1	1020	1290	35.9	28.4
2	980	1175	34.3	28.6
3	1215	1470	38.3	31.7
4	1140	1350	36.7	31.1
5	1070	1310	36.2	29.6
				149.4

$$\text{Avg. Ds} = 149.4/5$$
$$\text{Ds} = 29.9$$

Particle velocity versus scaled distance graph.   This method uses the scaled distance and the corresponding particle velocity for each shot as an ordered pair for plotting on a log-log graph. A series of test or production shots, a minimum of five, is used as before.

Scaled distance is the horizontal axis and particle velocity is the vertical axis. There must be a spread of data from low to high values, which can be achieved by placing the seismograph at a greater distance on each succeeding shot. If this is not done, the data will cluster together so that it will not be possible to draw the envelope curve.

Each ordered pair of scaled distance—particle velocity values is then plotted on a log-log graph. After the points are plotted, draw a straight line envelope so that all the points lie beneath the line. A reasonably accurate eye-ball fit is sufficient. If greater accuracy is called for, a best fitting regression line can be calculated and then the envelope line can be drawn parallel to it.

To determine a scaled distance value based on the test shots and the envelope, start on the particle velocity scale at the specified regulatory particle velocity, e.g., 1.0 in. per sec. Draw a horizontal line across the graph until it intersects the envelope line. At the point of intersection, drop a vertical line down to the scaled distance axis. The point at which it touches the scaled distance axis is the new value for the scaled distance. Blasting parameters e.g., charge weight or distance computed using this scaled distance should insure that particle velocities will be less than 1.0 in. per sec, providing that nothing in the system changes.

For other values of particle velocity e.g., 2.0 ips or 0.5 ips, start at the proper value and proceed in the same way. The following example illustrates the procedure:

Shot	D	W	$W^{0.5}$	Ds	PV
1	225	338	18.4	12.7	1.70
2	380	345	18.6	20.4	0.83
3	610	320	17.9	34.1	0.33
4	850	335	18.3	46.4	0.25
5	1060	405	20.1	52.7	0.16

The test data are plotted as a log-log graph because the propagation law

$$V = H \ (D/W^{0.5})^b \qquad (11.1)$$
$$V = H \ (Ds)^b \qquad (11.2)$$

reduces to a linear relationship when logarithms are taken on both sides on the equation.

$$\log V = \log [H \ (Ds)^b)] \qquad (11.3)$$

$$\log V = \log H + b \log Ds \qquad (11.4)$$

The scaled distance read from the graph is

$$Ds = 20$$

Charge weights and distances calculated using $Ds = 20$ should produce vibration levels less than 1.0 in. per sec.

Fig. 11.1(a) is a plotting of the scaled distance particle velocity data as it might be done at the operations site by the field engineer. This same data has been plotted and analyzed by a computer as shown in Figure 11.1(b). If computer analysis is available it would be quicker and more precise. The program used for this analysis is available from Precision Blasting Systems (1987).

As future blasts are fired and more data is available, it should be included with the test data. The addition of these new data points, should confirm the validity of the envelope line or indicate how it should be adjusted.

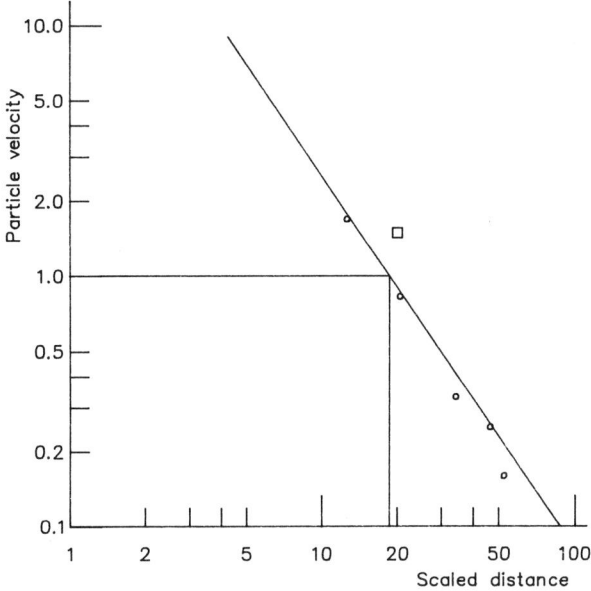

Figure 11.1(a)   Scaled distance versus particle velocity. (field drawing)

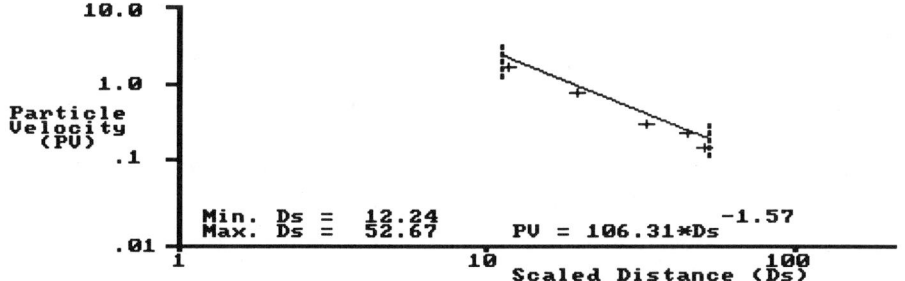

Figure 11.1(b)   Plot of data and 95% confidence-level line. (computer analysis)

When a shot goes bad as indicated by an excessive vibration level, it will show up by falling above the envelope line. This may be due to factors such as cap overlap, propagation, wet holes, poor drilling control and should be looked into to correct the condition. On Fig. 11.1(a) the data point surrounded by a square illustrates an out of control point based on two holes firing simultaneously instead of separately. It is good practice to add a safety factor to the adjusted Ds value. If the adjusted value is (23) then use a value of 27 or 29.

Scaled distance charts. Scaled distance charts can be constructed for any given values of the scaled distance. For example, choosing a scaled distance of 50, the explosive charge corresponding to specific values of the distance can be easily calculated. Since the log-log graph is linear—that is a straight line—it is only necessary to choose two distance values and compute the corresponding charges. The following illustrates the calculation for a scaled distance value of 50:

$$Ds = (D/W^{0.5}) = 50 \qquad (11.5)$$

$$W = (D/50)^2 \qquad (11.6)$$

Distance D (chosen) (ft)	Charge Weight W (calculated) (lb)
50	1
1000	400

By plotting these pairs of points (50, 1) and (1000, 400) on the log-log graph paper, and connecting them by a straight line, the result is the scaled distance curve for $Ds = 50$. Additional lines for scaled distances values of 10, 20, 60, 100, or any value, can be computed and plotted.

With these scale distance curves, it is possible to graphically determine the permissible explosive charge at any distance for a specified scaled distance value. Figure 11.2 is an example of a scaled distance chart for $Ds = 50$. The chart can be used in the following way. Choose a scaled distance of 50 as the operational level. The permissible charge at a distance of 600 feet can be determined as follows: Draw a vertical line upward from the distance value 600 until it intersects the $Ds = 50$ line. Then at this point of intersection on the scaled distance line, draw a horizontal line to the charge weight axis. This point is a value of charge weight, 144 lb, for the case in question.

Reversing this process will enable the operator to determine a compliance or safe value of the distance for a given charge weight of explosive and a specified scaled distance. For example, assume a charge weight of 225 lb and a scaled distance of 50.

Find the charge weight of 225 lb on the vertical axis at the left. Draw a line horizontally to the right until it intersects with the $Ds = 50$ line. At this point of intersection, drop a line vertically down to the distance scale of the bottom. The

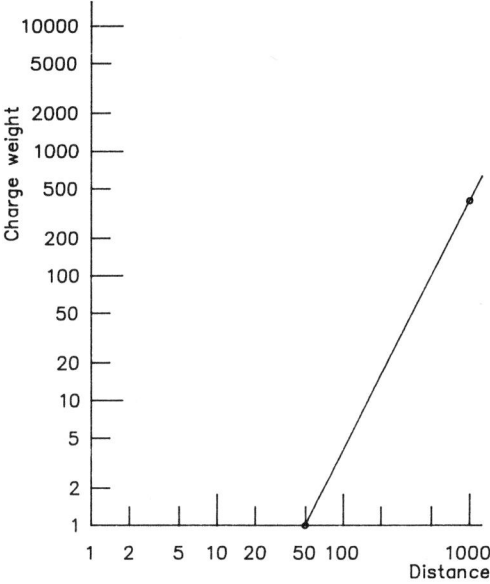

Figure 11.2   Scaled distance chart, $D_S = 50$.

point of intersection is the acceptable safe distance in this case, 750 ft. Obviously any distance greater than this will also be safe. Conversely, any distance less than 750 ft, i.e., 600 ft, is not in compliance and calls for an adjustment in procedure such as a reduction in charge weight.

Ground Calibration

When entering a new area in which the vibration conditions and seismic transmission characteristics are unknown, ground calibration can be done by firing a number of test shots.

Charge weight and distance are the principle factors that affect vibration and are subject to control. Other factors such as rock type, rock density, presence or absence of rock layering, slope of layers, nature of the terrain, blast hole conditions, presence or absence of water, also affect the transmission of vibration but are beyond control. The interaction of all these factors can be evaluated by firing several test shots and measuring the ground vibration. This is called ground or area calibration.

When the tests shots have been fired and recorded, a scaled distance-particle velocity plot on log-log graph paper is made. A minimum number of five shots will serve as a starter with more data to be added as additional shots are fired and recorded. All factors affecting vibration transmission are integrated together in this way. A safe working value for the scaled distance is determined from the log-log graph. All subsequent shots should generate vibration levels less than this particle velocity.

It is important and very useful to have an idea of the vibration level that corresponds to a given scaled distance value. The following table presents particle velocity corresponding to a specific scaled distance based on the modified Bureau of Mines equation.

$$V = 100 \ (Ds)^{-1.6} \tag{11.7}$$

Ds	V in./sec
10	2.5
11.5	2.0
20	0.83
30	0.43
40	0.27
50	0.19
60	0.14
70	0.11

Note the low particle velocities corresponding to scaled distance values of 50, 60, and 70. Ground calibration may indicate different particle velocities for a given area, maybe higher or perhaps lower, but this is what the operator needs to know in order to control the vibration and keep it safe.

Variability of Vibration

Vibration is a measured quantity. Its constituent elements charge weight, hole depth, density, and delay time are measured quantities and the other factors rock density, and rock layering terrain conditions are factors that cannot be controlled. The result is a scattering of data. If two blasts were identical with respect to charge, distance, rock formation, and so on—all measured or estimated quantities—the resulting vibration levels would most probably be different.

Scatter is a statistical property of a measured variable. If a large number of identical blasts were detonated and recorded by seismographs, at the same distance the recorded vibration levels will show a scatter that is distributed on both sides of an average value. This scatter is a characteristic factor in the data of the many investigations into the vibration problem.

This scatter of data approximates the normal distribution with a mean value and a standard deviation. This kind of data is usually grouped in convenient intervals and has the form shown in Fig. 11.3. Sigma, σ, is the standard deviation.

Understanding the nature of this scatter leads to the important conclusion that the scatter can be predicted and is subject to control. Higher than average vibration will occur regularly which is a cause for concern. Lower than average vibration will also occur, and while it is of no concern vibration wise it may be an indicator of a

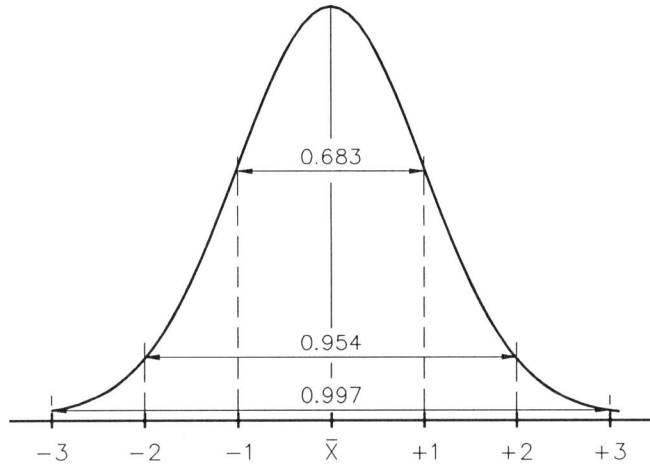

Figure 11.3 Standard normal distribution areas enclosed by successive standard deviations.

blast problem. Ninety-five percent of the data will lie within ± 2 σ of the mean, with half the data, or 47.5%, above the mean value and half below the mean value.

This scatter of data was analyzed for several different types of blasting operations: single hole construction shots, road construction shot, quarry shots, and strip mining coal shots. Based on this data the scatter in the vibration level is approximately 60% above and below the mean value when the blasting is apparently identical. For example, if the average vibration reading is 0.80 ips, readings could range from 0.32 to 1.28 ips. The scatter on the high side may result in exceeding a regulatory specification, or even damage to adjacent structures. Knowledge of the scatter potential allows for planning to prevent undesirable effects. A percentage (60%) of the mean was chosen for ease of field calculation and does not require calculation of the standard deviation.

Vibration standards.   The vibration problem came into being more than half a century ago. During the intervening years, many scientists and engineers investigated the problems and published their findings and the scope of the problem evolved. The first significant investigation was initiated by the U.S. Bureau of Mines in 1930, and produced Bulletin 442, Seismic Effects of Quarry Blasting in 1942. This and other programs will be briefly described.

Thoenen and Windes. "Seismic Effects of Quarry Blasting," U.S. Bureau of Mines, Bulletin 442, 1942.

Acceleration index
Safe zone   less than 0.1 g
Caution zone   between 0.1 and 1.0 g
Damage zone   greater than 1.0 g

Crandell, F.J. "Ground Vibration Due to Blasting and its Effect upon Structures," *Journal of the Boston Society of Civil Engineers*, 1949

Energy Ratio index $ER = (a/f)^2$

where

$$a = \text{acceleration in ft/sec}$$
$$f = \text{frequency in Hz}$$

Safe zone $= ER$ less than 3
Caution zone $= ER$ between 3 and 6
Damage zone $= ER$ greater than 6

Energy ratio has the dimension of velocity and an $ER = 1$ is equivalent to a particle velocity $= 1.9$ in./sec

Langefors, Westerberg, and Kihlstrom. "Ground Vibration in Blasting," Parts I–III, Water Power, 1958.

Velocity index
No damage   less than 2.8 in./sec
Fine cracks   4.3 in./sec
Cracking   6.3 in./sec
Serious cracking   9.1 in./sec

Edwards and Northwood. "Experimental Blasting Studies on Structures," National Research Council. Ottawa: Canada, 1959.

Velocity index
Safe zone   less than 2.0 in./sec
Damage   4.0 to 5.0 in./sec

Nichols, Johnson, and Duvall, "Blasting Vibration and Their Effects on Structures," U.S. Bureau of Mines, Bulletin, 656, 1971.

Velocity index
Safe zone   less than 2.0 in./sec
Damage zone   greater than 2.0 in./sec

The Bureau of Mines combined its work with that of other important investigations and in Bulletin 656 presented the results listed above. Particle velocity is considered to be the best measure of damage potential and the safe vibration criterion was specified in Bulletin 656 as follows:

The safe vibration criterion is based on the measurement of individual components, and if the particle velocity of any component exceeds 2.0 in./sec, damage is likely to occur.

The concept of damage is critical and in Bulletin 656 damage means the development of cracks in plaster. Since the particle velocity was specified as the measure of damage, it also became the measure of safety. If damage is likely to occur above 2.0 in. per sec then below 2.0 is safe, so the reasoning went, and 2.0 in. per sec became known as the safe limit. Many regulations were and still continue to be based on this value. Some additional levels of vibration based on the results of other investigations used in Bulletin 656 are as follows:

Threshold of damage (4.0 in./sec)
      opening of old cracks
      formation of new cracks
      dislodging of loose objects

Minor damage (5.4 in./sec)
      fallen plaster
      broken windows
      fine cracks in masonry
      no weakening of structure

Major damage (7.6 in./sec)
      large cracks in masonry
      shifting of foundation-bearing walls
      serious weakening of structure

The major damage category in compatible with the beginning damage level for natural earthquakes when frequency of vibration is taken into consideration. The frequency spectrum of natural earthquakes is much different than that of blasting vibration.

Recent damage criteria.   The U.S. Bureau of Mines Report of Investigation RI 8507, (Siskind, et. al) documented an investigation of surface mine blasting. Structural resonance lies in the low frequency range typically 5–20 Hz. Blast vibration in this frequency range can cause a resonance response in structures which produces increased displacement and strain. This was found to be a serious problem in RI 8507.

Frequency was now an important factor in assessing the damage potential of vibration. The Bureau of Mines also made extensive measurements inside structures. This seems like a proper approach since that is where the problem exists, cracks in plaster. Early investigations of structural vibration, however, yielded very poor results; hence, the emphasis on ground measurement and away from structural measurement.

RI 8507 specified a threshold of damage as "cosmetic damage of the most superficial type, of interior cracking that develops in all homes, independent of blasting." The safe vibration level was defined as levels unlikely to produce interior cracking or other damages in residences.

The safe vibration levels specified in RI 8507 are shown in Fig. 11.4. These are based on a 5% probability of damage.

These safe vibration levels represent a conservative approach to vibration

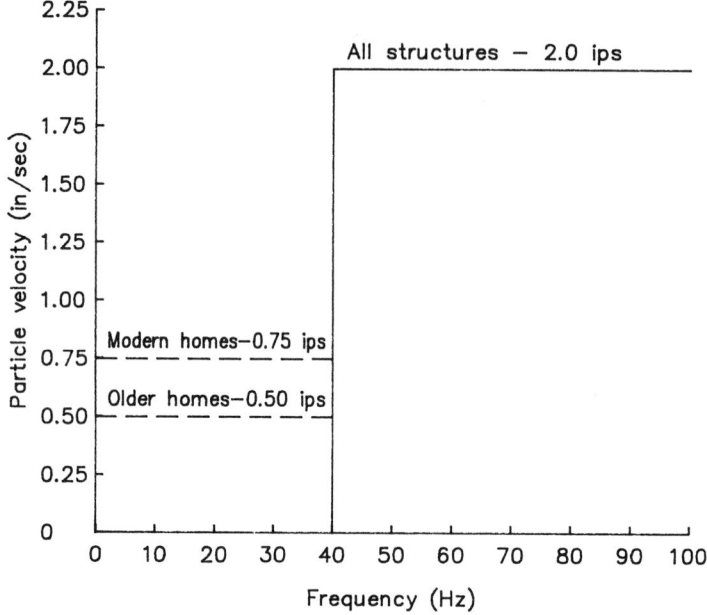

Figure 11.4   Safe vibration levels. (From RI 8507, U.S. Bureau of Mines.)

damage control. Little or no damage is likely to occur when they are in force. However, the increase in damage would probably not be significant if a simple 1.0 in. per second regulation were applied across the vibration frequency spectrum.

Alternative Blasting Criteria

In addition to the above proposed vibration standards an "Alternative Blasting Level Criteria" was put forth in RI 8507. This alternative blasting level criteria employed displacement and velocity each applied over specific frequency ranges. The measurement requirements are much more severe than anything used up to this time. Fig. 11.5 shows their criteria.

When these criteria using both displacement and velocity over respective frequency ranges were presented there was no instrumentation available that could do this. This gave an impetus to the industry and soon instrumentation capable of measuring displacement, velocity, acceleration, frequency, and resultant particle velocity became available.

In RI 8896, (1984), "Effects of Repeated Blasting on a Wood-Frame House" U.S. Bureau of Mines, it states that cosmetic cracks occurred during construction of the test house and also during periods when no blasts were detonated. Cosmetic crack formation increased when ground particle velocity exceeded 1.0 in./sec. It was further noticed that human activity, temperature, and humidity changes caused strains equivalent to ground particle velocity of 1.2 in./sec to 3.0 in./sec.

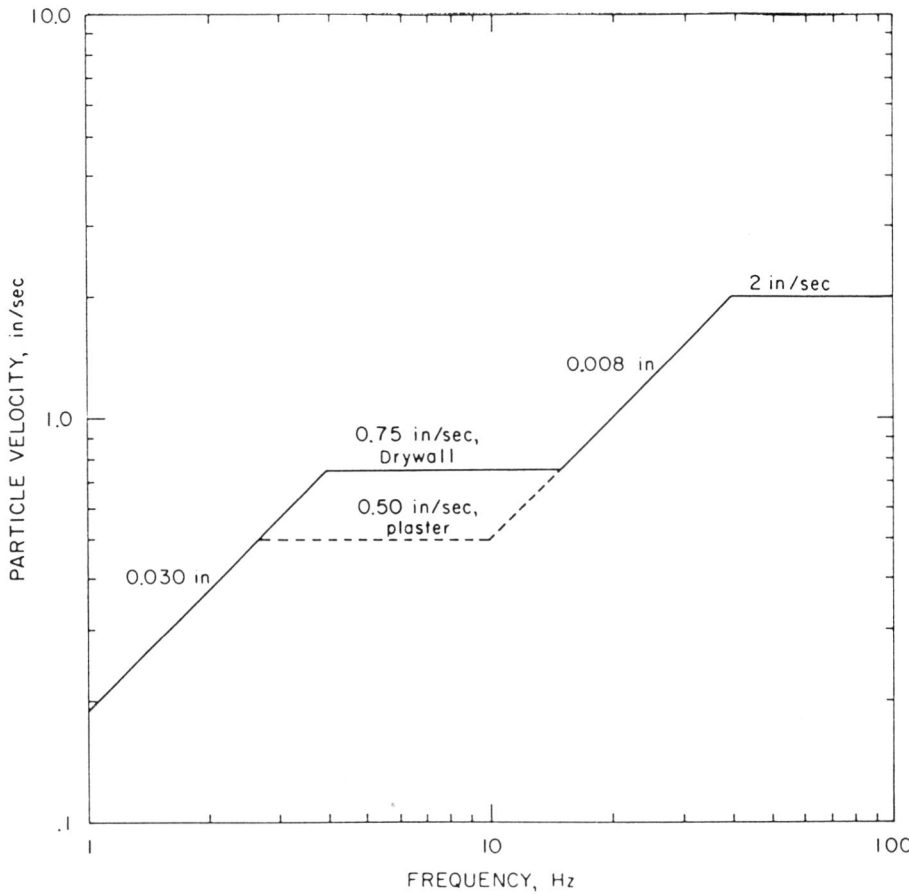

Figure 11.5 Alternative blasting-level criteria. (From RI 8507, U.S. Bureau of Mines.)

What constitutes vibration damage? This is not a simple question. The concept of damage has changed from cracks in plaster—readily visible cracks—as presented in Bulletin 442 and Bulletin 656 to "cosmetic damage of the most superficial type . . ." in RI 8507. Most structures have cracks of some kind usually due to normal environmental stresses. When blasting comes into the area it usually gets credit for these environmentally induced cracks. Vibration damage because of blasting can and does occur but blasting operations are so well engineered and controlled today that genuine cases of blasting damage are a small minority. Annually millions of blasts are detonated that are safe and do not cause damage.

It is not likely that a lower limit of vibration exists beyond which damage will not occur. There will always be structures at the point of failure due to normal environmental stresses waiting for the proverbial straw that broke the camel's back. That straw may be vibration but in many cases there is no apparent cause and an intensive investigation is usually necessary to ascertain the cause. There are many reported cases of structural collapse, apartment building, commercial buildings, sports arenas, and so on, which were unexpected and had no apparent cause. Had a vibration source suddenly appeared on the scene e.g., construction blasting, drop ball operation, demolition, it undoubtedly would be judged to be the cause when in truth the structure was ready to collapse, an accident waiting to happen.

The Office of Surface Mining Regulations

The Office of Surface Mining has adopted a modification of the Bureau of Mines Alternative Safe Blasting Criteria for the surface mining industry. Since RI 8507 proposed that frequency is an important factor affecting vibration as well as distance, the Office of Surface Mining presented its regulation as shown in Table 11.1.

It may seem strange that a higher particle velocity is permissible at short distance than at greater distances. This reflects two things: First high-frequency vibration attenuates more rapidly with distance than low-frequency vibration, hence, low-frequency vibration persists longer. Second, structural resonance response associated with the low-frequency vibration may occur.

At short distances, high-frequency vibration predominates. At larger distances, the high-frequency vibration has attenuated or died out and low-frequency vibration predominates. Buildings have low-frequency response characteristics. If the structural frequency and the ground frequency match, resonance can occur and at high particle velocity damage may result. Therefore, at large distances a lower peak particle velocity, 0.75 ips, and a larger scaled distance, $Ds = 65$, are mandated. At the shorter distances, a higher peak particle velocity, 1.25 ips and a smaller scaled distance, $Ds = 50$, are permitted.

The Office of Surface Mining graph for displacement and velocity and the frequency ranges over which each applies are shown in Figure 11.6. Note that the 2.0 in./sec range begins at 30 Hz as distinct from the USBM RI 8507 range which begins at 40 Hz.

Regulation enforcement.   On the job regulation of vibration and general safety measures is generally relegated to the safety forces, fire department, police department or there may be a person specifically responsible for this job.

TABLE 11.1   Office of Surface Mining Required Ground Vibration Limits

Distance (D) from the blasting site (ft)	Maximum Allowable Peak Particle Velocity (V max) for Ground Vibration (in./sec)	Scaled-distance Factor to be Applied Without Seismic Monitoring
0–300	1.25	50
301–5000	1.00	55
5001 and beyond	0.75	65

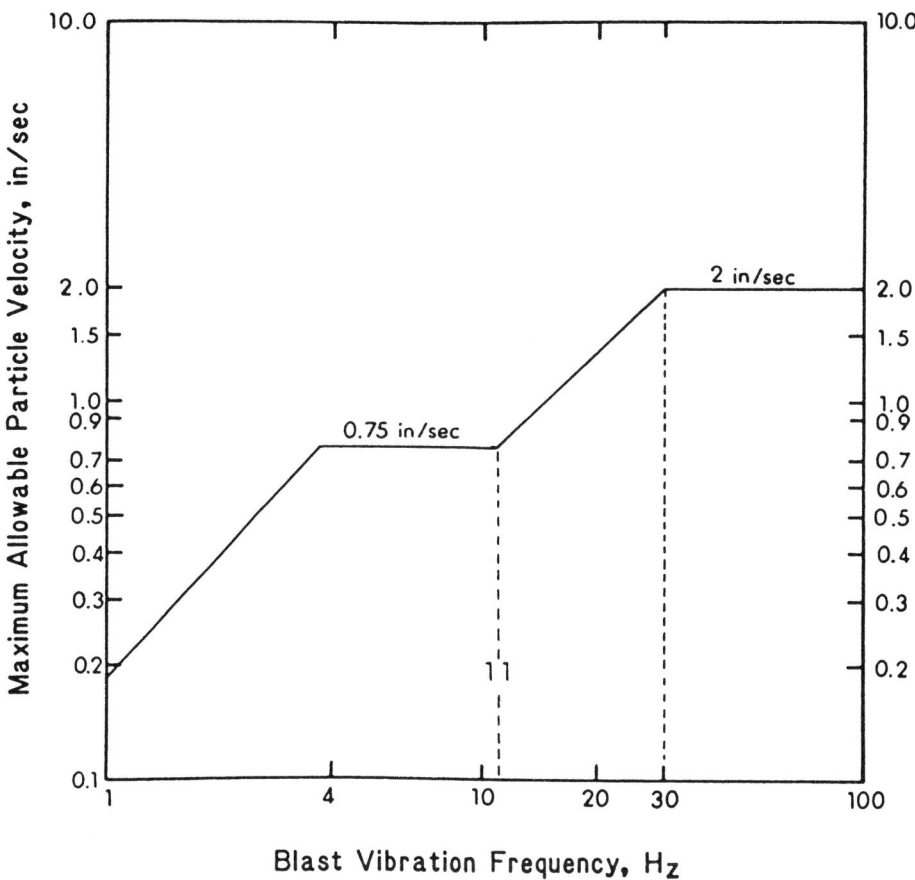

Figure 11.6 OSM alternative blasting-level criteria. (Modified from Fig. B1, RI 8507, U.S. Bureau of Mines.)

Procedures will vary with the type of job, is it governmental, federal, state, county, municipal, or is it a private industry job? Some political entities such as municipalities with previous experience have regulations based on 2.0 in. per sec peak particle velocity , others use 1.0 in. per sec peak particle velocity, while still others leave it up to the discretion of a hired consultant. Surface mining operations are generally governed by the Office of Surface Mining Regulations.

Reports must be made for all blasts and copies must be filed with the principals involved. In the case of public projects, they become part of the public record that is available for examination and study by interested parties. In the case of private industry, the reports are usually proprietary.

Not uncommonly, an initial blasting plan is required before blasting permits are issued. Significant variations from this plan must be requested in writing by submission of the modified plan and justified before approval is granted.

In the case of violation of the blasting regulations, work may be stopped with start up delayed until the violation is accounted for and an adjusted blasting plan presented that will avoid a repeat of the violation.

Characteristic Vibration Frequencies

Different types of blasting generate different frequencies of vibration. Characteristic frequencies associated with coal mine blasting, quarry blasting, and construction blasting are illustrated by the U.S. Bureau of Mines in R.I. 8507. The lowest frequencies are associated with coal mine blasting, intermediate frequencies with quarry blasting, and high frequencies with construction blasting. Figure 11.7 presents this data.

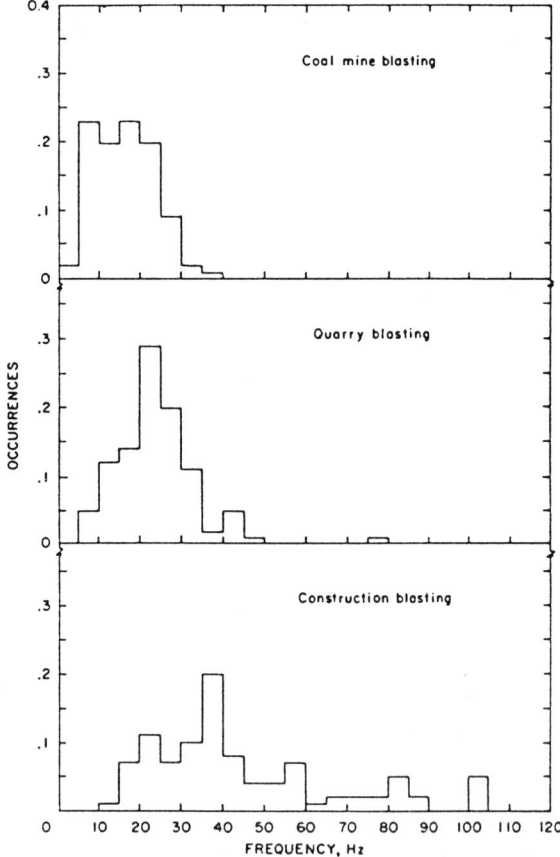

Figure 11.7 Frequencies from coal mine, quarry and construction blasting. (From RI 8507, U.S. Bureau of Mines.)

This description of frequency association is somewhat interpretative. The differences are due to physical parameters such as shot size, hole diameter, distance and rock properties, all interacting when a blast is fired. Distance is probably the most important control factor since vibration attenuates as the square of the frequency. As previously noted, the high-frequency vibration predominates at short distances and low-frequency at larger distance. Hence, low-frequency vibration will appear on any blast record if the recording distance is sufficiently large.

Spectral analysis.   A vibration record is a complex of many vibration frequencies, some of which are more important and some less important. Spectral analysis is a method for analyzing the frequency content of a vibration record and the relative importance of each frequency. This is a significant advance over other methods for determining the effects of frequency. The standard vibration record of a blast shows how the ground motion changes with time and is referred to as a time-domain record. The record is digitized, usually at 1 ms intervals, and the digitized data are subjected to a Fourier Analysis. This shows the relative vibration levels associated with each frequency. The data is now said to be in the frequency domain.

Figure 11.8 illustrates this process. To the left, is the time domain or standard record. To the right is the frequency domain record generated by the Fourier Analysis. This is taken from RI 8507, Siskind, et. al, 1980.

Response spectra.   Response spectra is a mathematical technique that predicts the response of the structure to a given vibration. It was previously noted that coal mine blasting, quarry blasting, and construction blasting, each show characteristic frequencies of vibration. Coal mine blasting generates the lowest frequencies and these become a problem because of structural resonance. A given structure with its own response characteristics will, therefore, respond differently to each of these different frequency generating blasts. Structures also differ, so that two structures may respond differently to the same blast.

Response spectra analysis considers a structure to be a damped oscillator, with its own natural frequency of vibration. The equation of motion of this damped oscillator is programmed into a computer. The digitized blast vibration data are fed into the computer (impressed on the structure) which calculates the structural response or displacement for this blast. The assumed natural frequency of the struc-

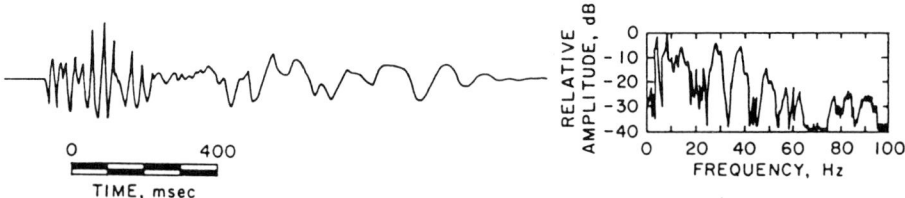

Figure 11.8   Coal mine blast vibration record and frequency spectrum. (From RI 8507, U.S. Bureau of Mines.)

ture, and the maximum displacement calculated by the computer are an ordered pair and form one point on the response-spectra curve.

The process is repeated for additional assumed natural frequencies of the structure and each assumed natural frequency with the calculated maximum displacement is an additional point for the response spectra curve. When all the assumed natural frequencies, usually 1 to 100 Hz, and the corresponding calculated maximum displacements have been plotted and the points joined together, the result is the response-spectra curve. What does the response-spectra curve tell the analyst? It says something like this: If the structure has a natural response frequency of 7 Hz and a blast with 7 Hz frequency causes it to vibrate it will be displaced 0.03 in. Similarly for any other frequency. Note that the response-spectra gives relative displacement values. If relative velocity data is required it can be computed by multiplying the displacement by $2\pi f$, where f is the assumed natural frequency of the structure. Using the previous example in which the response-spectra curve gave a displacement of 0.03 in. at a frequency of 7 Hz the relative velocity is

$$V = 2\pi f D$$
$$V = 2\pi\, 7 \times 0.03 \tag{11.8}$$
$$V = 1.32 \text{ in. per sec}$$

A relative maximum velocity response curve can be calculated in this way. The relative maximum velocity is termed maximum pseudo velocity.

Using response spectrum analysis, one can estimate the response of a structure to an actual blast vibration, thus anticipating, and hopefully eliminating problems before they arise. The implications for structures such as nuclear power plants is obvious.

Response-spectra analysis of day to day blast vibration is available. Vibra-Tech Engineers of Hazelton, PA, have developed a program referred to as RSVP, Response-Spectra Velocity Profile. A standard analog vibration record of a blast provides the basic data for the RSVP. The data is digitized and a response-spectra velocity curve is calculated over a frequency range 1 to 100 Hz. The amplification of the vibration in the structure due to resonance is quickly apparent. The amplification may range as high as 5 to 6 so that a ground particle velocity of 0.25 in. per sec amplified in the structure by a factor of 5 becomes 1.25 in. per sec in the structure. If the blast vibration is in the high-frequency range beyond the normal resonance range for residential structures 5–20 Hz then the house vibration will not be amplified but may actually be deamplified. This also shows the importance of avoiding low-frequency vibration when possible.

By using response-spectra analysis the vibration of a structure can be anticipated and thus, problems can be avoided before they occur.

Long term vibration and fatigue. Blasting vibration is a short term transient phenomenon usually lasting a second or two frequently less. The question of repeated blasting effects arises regularly as a point of concern such as: it didn't damage my house this time, but what will happen if it keeps up? The effects from pile driving and recurring industrial operations are of similar concern. Generally, these vibration levels are relatively low with low damage potential.

Much less information is available on this topic than on blasting in general and it is not regarded as an urgent problem.

Walter (1967) investigated the effect of impact vibration on a structure. Vibration continued for 13 months, 24 hours a day. The test structure was an ordinary room approximately 8 by 8 by 8 ft of dry wall construction. The vibrator was mounted on the ceiling.

The wall panels had a natural frequency of 12.5 Hz and the ceiling panel was 60 Hz. Vibration frequencies measured in the wall panels ranged from 10 to 18 Hz with particle velocity ranging from 0.05 to 0.16 in. per sec.

In spite of the long time involved, approximately thirty million seconds and the many impacts imparted to the structure, no damage occurred. Inspection of the structure revealed no observable effects. It was concluded that low level vibration even in the natural frequency response range of the structure has little potential for causing damage.

An extensive fatigue test was conducted for the Bureau of Mines by the U.S. Army Corp. of Engineers, Civil Engineering Research Laboratory, CERL. An 8 by 8 by 8 ft test room was mounted on a biaxial shake table which was programmed with one horizontal component and the vertical component of a quarry blast from Bulletin 656. The predominant blast frequencies were 26 and 30 Hz respectively.

Vibration tests were run at 0.1, 0.5, 1.0, 2.0, 4.0, 8.0, and 16.0 in. per sec. Each was run a series of times starting with 1 run, then 5 runs, then 10, 50, 100, and 500 runs with inspection after each series. No damage occurred until the sixth run at 4.0 ips. This sixth run was preceded by 2669 prior runs with no damage. In fact, there were 666 runs at 2.0 ips and 5 at 4.0 ips with no damage. It is significant to note that when damage occurred it occurred at a particle velocity in excess of 2.0 ips.

Resonant frequency test of 1/10 scale block masonry walls were conducted by Koerner. Failure was observed after approximately 10,000 cycles at particle velocities of 1.2 to 2.0 ips. Additional tests on 1/4 scale block walls showed cracking after 60,000 to 400,000 cycles at particle velocities 1.69 to 1.95 ips.

The investigation RI 8896 by the Bureau of Mines tested a wood frame house by subjecting it to 587 production blasts after which it was shaken mechanically to determine the threshold of fatigue cracking. The first crack appeared after 56,000 cycles, which is the equivalent of 28 years of blasting twice a day at a vibration level of 0.5 in/sec the damage was a cracked wall-board tape joint.

These studies show that fatigue effects such as cracking may occur at vibration levels that are relatively high, in fact, there would be concern about the damage potential of a single vibration event of this level.

Vibration effects.   Natural earthquakes produce a characteristic crack pattern called the X-crack. This is a relatively low earthquake intensity effect. These cracks result from the fact that the base of a structure, moves with the vibrating earth while the top of the structure lags behind because of its inertia. The structure is deformed from a rectangular shape into a parallelogram, with one of its diagonals elongated and the other compressed. If the elongation exceeds the strength of the material, a tension crack results. When the vibration reverses, the opposite diagonals will be elongated and compressed with the possible formation of another tension crack. When both cracks occur, they form an X-pattern, or X-crack. Figure 11.9 illustrates the process.

Blasts that produce large displacements and large strains are most likely to cause X-cracks if they occur and large displacement and strain are likely to be associated with low frequency vibration. The pattern is not a strong identifying characteristic of blast damage.

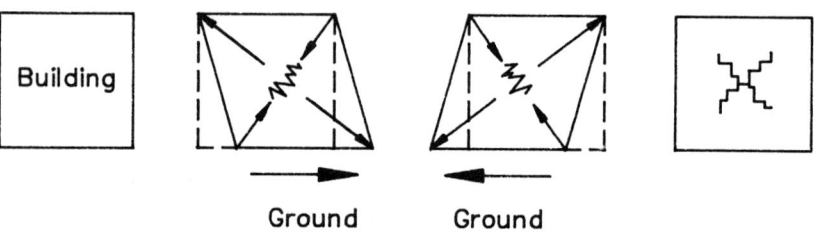

Figure 11.9   Vibration X-crack pattern.

Non-damage effects.   Blasting is now an engineered science and controlled blasting is the normal procedure. Damage-producing vibration seldom occurs. However, many other effects occur that are disconcerting and alarming to persons who feel and hear the vibration. Some of these effects are

> walls and floors vibrate and make noise
>
> pipes and duct work may rattle
>
> loose objects, plates, etc., may rattle
>
> objects may slide over a table or shelf, and may fall off after repeated shaking
>
> chandeliers and hanging objects may swing
>
> water may ripple and oscillate
>
> noise inside a structure is greatly amplified over noise outside
>
> vibration is very disturbing to occupants.

Causes for cracks other than blasting.   A house without cracks would be unusual. Cracks occur in the walls and ceilings of structures, and there are many causes ranging from poor construction to normal environmental stress, such as thermal stresses, humidity, wind, and so on. *The Small Home*, published by the Architects' Small House Service Bureau of the United States, Inc., 1925,* gave a list of reasons for the development of cracks, which included the following:

Building a house on a hill

Failure to make the footings wide enough

Failure to carry the footings below the frost line

Width of footings not made proportional to the loads they carry

The posts in the basement not provided with separate footings

Failure to provide a base raised above the basement floor line for the setting of wooden posts

Not enough cement used in the concrete

Dirty sand or gravel used in the concrete

Failure to protect beams and sills from rotting through dampness

Setting floor joists one end on masonry and the other end on wood

Wooden beams used to support masonry over openings

Mortar, plaster, or concrete work allowed to freeze before setting

Braces omitted in wooden walls

Sheathing omitted in wooden walls (excepting in "back-plastered" construction)

Drainage water from roof not carried away from foundations

Floor joists too light

Floor joists not bridged

Supporting posts too small

Cross beams' too light

Subflooring omitted

Wooden walls not framed so as to equalize shrinkage

Poor materials used in plaster

Plaster applied too thin

Lath placed too close together

Lath run behind studs at corners

Metal reinforcement omitted in plaster at corners

Metal reinforcement omitted where wooden walls join masonry

Metal lath omitted on wide expanses of ceiling

Plaster applied directly on masonry at chimney stack

Plaster applied on lath that are too dry

Too much cement in the stucco

Stucco not kept wet until set

Subsoil drainage not carried away from walls

First coat of plaster not properly keyed to backing

Floor joists placed too far apart

Wood beams spanned too long between posts

Failure to use double joists under unsupported partitions

Too few nails used

Rafters too light or too far apart

Failure to erect trusses over wide wooden openings

Reduction of Vibration

If for any reason the vibration level is getting high or near a regulatory limit, it would be prudent to give some consideration as to how to reduce it.

Charge reduction. Reduce the number of holes per delay. If it is already down to one hole per delay then try decking holes and fire each hole with two or more delays.

Blast design. It may be necessary to redesign the blast so that less energy per hole is necessary to fragment the rock. It may be necessary to change the hole spacing, the burden and even the hole size. A change in explosive may be helpful also. This is an extreme circumstance and hopefully will not occur.

Seismic methods. Several methods of vibration reduction have been developed that use seismic data as the basis. The Pre-Seis blast simulation technique discussed in Chap. 8 takes into account blasting parameters, blast geometry, and tolerances. The Pre-Seis model is based on sound engineering principles. Other seismic methods are also available that generate a synthetic seismogram using the wave signal or signature of a single shot.

By means of a computer, a multiple hole shot is simulated by feeding the digitized wave signature data into the computer at the specified delay intervals. To illustrate, assume a four-hole shot with four delays, 25 ms, 50 ms, 75 ms, and 100 ms. The data is fed into the computer at these delay times and the combined motions produce a composite wave form with specific frequencies and peak particle velocities.

The composite motion or synthetic seismogram is then analyzed for its frequency content and particle velocity to see which frequencies produce the larger particle velocities and most importantly which delay intervals reduce the peak particle velocity.

In practice, the delay interval may be varied by 2 ms time intervals over any desired range and from this the optimum firing intervals or delays can be specified for a particular shot.

This methodology is a significant step in vibration control but it has severe

*Published in Monthly Service Bulletin 44 of the Architects' Small House Service Bureau of the United States, Inc.

limitations. It assumes a point source rather than the actual blast geometry. Cap scatter has not been addressed. No blasting pattern variables are controlled. The application is site specific which means that all the factors usually affecting a shot, local geology, rock layering, rock type terrain, direction, and so on also affect this technique. These will change with direction, location, and area so that any change represents new conditions. Keep in mind that this methodology is essentially addressing delay timing and how best to effectively use the delays.

Blasting standards for nonresidential structures.   Other structures exist that may require special consideration when exposed to blast vibration. These can be divided into two groups, high-level vibration structures, and low-level vibration sensitive components.

Green concrete, cured concrete.   Green concrete is in the low-level group. The period between 4 and 24 hr is particularly critical when crystallization is beginning. Concrete acquires about one-third its strength in 72 hr. Before 72 hr blasting is not advised. After this time a peak particle velocity of 1.0 in. per sec may be used until 7 days after which 2.0 in. per sec may be used. Concrete reaches full strength at 28 days. Well cured concrete should be able to withstand 5.0 in. per sec.

Bridges.   A steel bridge or a reinforced concrete structure of integrity would minimally be covered by 2.0 ips and probably go to 5.0 ips. A wooden structure should withstand 2.0 in. per sec.

Buried pipe lines.   Buried pipelines such as gas and oil transmission lines are high-strength structures and can withstand high-peak particle velocity. However, the pipe should be in the elastic zone and never in the fracture zone. If the rock is fractured it may fracture whatever is buried in it. To stay out of the fracture zone use a stand off distance from the blasthole equal to 3 to 5 times the hole spacing. For a hole spacing of 10 ft, the stand off distance is 30 to 50 ft. A particle velocity of 5.0 ips should be safe.

Computers and hospitals.   Computers and hospitals fall into the low-level sensitive category. Hospital equipment is usually the point of concern not the hospital structure and usually there are no vibration specifications for it. It simply cannot be shaken according to the hospital personnel. A practical procedure is to measure the ambient background vibration in the sensitive areas of the hospital and compare this with a test shot. People are usually surprised to learn how much vibration occurs normally that they are not aware of.

Computer specifications.   Computer specifications are usually frequency dependent having different vibration levels for different frequency ranges. One computer manufacturer has the following specifications (Table 11.2):

TABLE 11.2 Vibration Specifications for Computer System

Frequency Hz	Double Amplitude	Acceleration
5–25	0.001 in. (0.025 mm)	
25–100	0.0005 in. (0.013 mm)	
100–300		0.25G (2.45 m/s^2)

Sensitivity to Vibration

Blasting today is highly technological, precisely engineered, well regulated if not over regulated with the result that blasting is safe, in fact, very safe. In spite of this, there are numerous complaints. Why? People are very sensitive to vibration. If this were not so, the vibration problem would scarcely exist. The explosives technology of today insures that most operations are conducted in a safe manner with a very low probability of damage.

When a person feels a blast vibration the reaction may range from one of curiosity, to concern, even to fear. This is human response and it is important to know something about how humans respond to vibration. Physical factors such as vibration level, frequency, and duration all play a part. Human factors such as age, health, and environment are also important and contribute to the subjective character of human response to vibration.

An early investigation of human response was conducted by Reiher and Meister, Berlin, 1931. Various other investigations by Goldman, 1948, and Wiss and Parmelee, 1974 contributed to our knowledge. The results of these investigations was presented graphically in the U.S. Bureau of Mines RI 8507, Siskind, et al, 1980 (Fig. 11.10).

The human response curves are very similar. They are also highly subjective, blending the physiological and psychological sensitivity of each person. Using this data, a simple set of human response criteria has been developed. This is shown in Table 11.3.

Table 11.3 illustrates the factor of 100 rule of thumb. If 2.0 in. per sec is taken as the damage level then vibration is easily noticeable to people at 1% of this or conversely when people just feel the vibration it would have to be approximately 100 times more intense to reach the damage level.

Because of this acute sensitivity people commonly have extreme reaction to vibrations and tend to describe them in much exaggerated terms. For example, one

TABLE 11.3 Simplified Human Responses

Response	Particle Velocity ips.	Displacement in inches at 10 Hz	at 40 Hz
Noticeable	0.02	.00032	.00008
Troublesome	0.20	.0032	.0008
Severe	0.70	.011	.0028

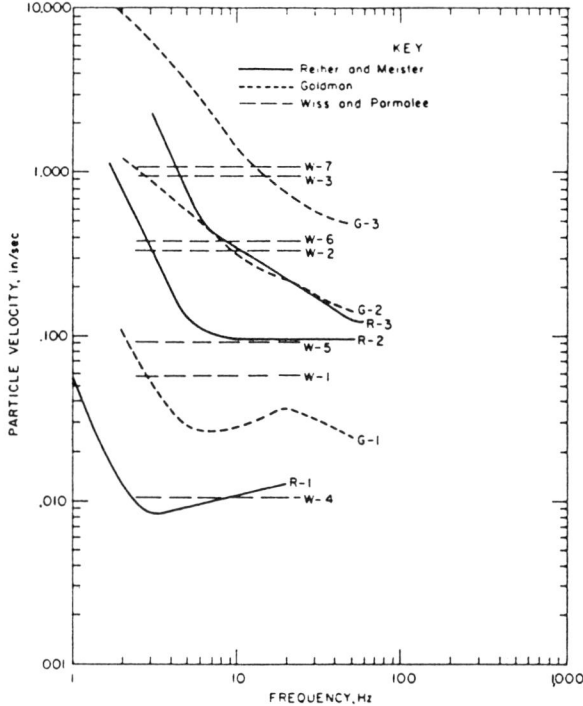

Figure 11.10   Human response to vibration. (From RI 8507, U.S. Bureau of Mines.)

individual stated "you're wrecking my home." When informed that the seismograph registered a low level of vibration, completely safe for residential structures he replied, "well if you keep it up you will wreck it."

Cultural and acultural vibration.   People are constantly exposed to vibration in one form or another but are seldom aware of it. Such vibration has been designated cultural vibration. Generally, it elicits no reaction from the person affected.

Other vibration effects us quite differently. We are aware of it, we usually do not like it, it is not part of the daily experience, it is unusual. This has been designated acultural vibration.

Some example of cultural and acultural vibration are listed in the following table:

Cultural Vibration	Acultural Vibration
Automobile	Blasting
Commuter train	Pile driving
Household appliances	Impact machinery
Industrial operations	Forging operations
Airplane	Jack hammer

For the average person blasting is the most acultural experience. There is annoyance and fear associated with it because explosives are associated with violence, destruction, and war and generate fear. All this, however, occurs at a level well below the level necessary to cause damage to structures.

Air Blast Monitoring and Control

Airblast.   When a blast is fired it frequently is accompanied by a loud noise called airblast. Airblast, however, is not simply the sound that is heard. Airblast is an atmospheric pressure wave consisting of high-frequency sound that is audible and low-frequency sound or concussion that is subaudible and cannot be heard. Either or both of the sound waves can cause damage if the sound pressure is high enough. Airblast, generally, is an annoyance problem, does not cause damage but may result in confrontation between the operator and those affected.

Airblast is generated, in several ways, by the explosive gases being vented to the atmosphere as the rock ruptures, by stemming blowout, by displacement of the rock face, by displacement around the bore hole and by ground vibration. Various combinations may exist for any given blast.

Overpressure and decibels.   Airblast overpressure is measured in decibels (dB) or in overpressure in pounds per square inch (psi). Because of the wide range of overpressure, more than one billion units, the decibel is usually used. The decibel is defined in terms of the overpressure, by the Eq. 11.9.

$$dB = 20 \log (P/P_o) \tag{11.9}$$

where

dB = Sound level in decibels
P = Overpressure in psi
P_o = Overpressure of the lowest sound that can be heard
$P_o = 3 \times 10^{-9}$ psi

Some typical sound levels with values in both dB and psi are shown in Fig. 11.11.

Different weighting networks are used to measure sound levels depending on the goal to be achieved. The networks are designated A, B, C, and Linear and differ essentially in the ability to measure low-frequency sound. The A-network corresponds most closely to the human ear and discriminates severely against the low frequencies. The B-network discriminates moderately and the C-network only slightly while the Linear network measures all frequencies.

Blast sound is primarily in the low frequency range, and a sound measuring device must have a low frequency response capability to accurately measure the sound level intensities. The linear peak network is preferred for blasting work although the C-weighted network is sometimes used.

Spectral analysis of blast sounds were done by Siskind and Summers, 1974, which clearly showed the very low subaudible frequencies.

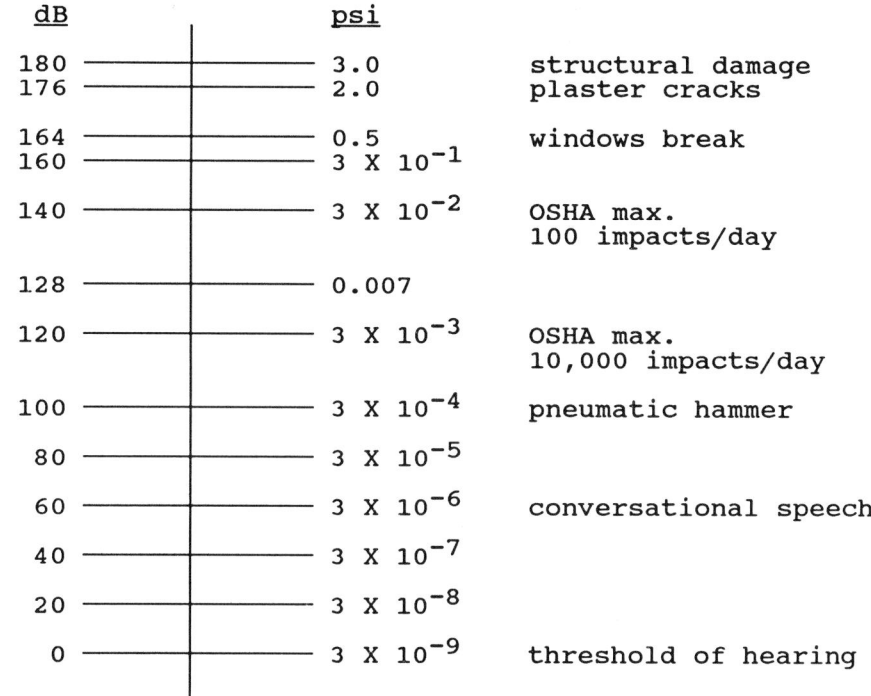

Figure 11.11   Typical sound levels.

Glass Breakage

Bulletin 656 presented the results of extensive testing by the U.S. Bureau of Mines to determine the sound levels likely to cause glass breakage, and the scaling law that would apply. Glass breakage occurs at much lower levels of overpressure than structural damage, such as cracking plaster. Thus, the absence of glass breakage precludes structural damage. Air blast regulation is keyed to glass breakage.

An overpressure of 0.5 psi. (164 dB) was proposed in Bulletin 656 as a safe level to prevent glass breakage. It further indicated that if ground vibration were below 2.0 in./sec then airblast overpressure was automatically limited to safe levels, that is, less than 0.5 psi. (164 dB).

Bureau of Mines TPR 78 (1974) Siskind and et al. proposed lower safe levels for preventing glass breakage. These levels also were aimed at reducing annoyance resulting from the airblast. These values are shown in Table 11.4.

Scaled distance for airblast.   Airblast is scaled according to the cube root of the charge weight, that is

$$D/W^{0.33} = K \tag{11.10}$$

TABLE 11.4 Sound Level Limits

	Sound Level Limits			
	Linear peak		C-peak or C-fast	A-peak or A-fast
	dB	psi	dB	dB
Safe	128	0.007	120	95
Caution	128 to 136	0.007 to 0.018	120 to 130	95 to 115
Limit	136	0.018	130	115
	Recommended		Not Recommended	

(U.S. Bureau of Mines TPR 78, 1978).

where

$$D = \text{distance in ft}$$
$$W = \text{maximum charge per delay in lb}$$
$$K = \text{scaled distance value}$$

Vibration, however, is scaled according to the square root of the charge

$$D/W^{0.5} = K \tag{11.11}$$

Siskind and et al. TPR 78 (1974) proposed .007 psi as the safe overpressure for glass. Interpolating the airblast scaled distance diagram of Bulletin 656 using .007 psi give an approximate value for $K = 180$, or

$$D/W^{0.33} = 180 \tag{11.12}$$

There are obvious inconsistencies between the scaled distance formulas for vibration and for glass breakage.

vibration $D/W^{0.5} = 50$
                for $D = 50$ ft $W = 1$ lb of explosive

glass breakage $D/W^{0.33} = 180$
                for $D = 50$ ft $W = 0.021$ lbs of explosive

What this says is that at 50 ft, glass safety would allow only 0.021 lb of explosive to be shot but plaster safety (no cracks) would allow 1.0 lb of explosive. When blasting is conducted safely for plaster cracks, there is no glass breakage.

How to resolve this? Pay no attention to the factor $K = 180$. Do not use it. Measure the blast overpressure with a linear scale sound level meter. Blast sound in the near field (close by) under normal conditions has an approximate range 114 to

124 dBA. The .007 psi corresponds to 128 dB. If the sound level begins to crowd into the 128 dB range, institute corrective measures.

These inconsistencies probably are due to an admixture of conservative physical parameters with psychological annoyance factors.

Cube root scaling, however, is appropriate for airblast, that is, the observed data conform to cube root scaling rather then square root scaling. An example of airblast data in psi and cube root scaling from Bulletin 656, U.S. Bureau of Mines is shown in Fig. 11.12. Computer Programs are available to do regression analysis on airblast data and produce graphs similar to Figure 11.12.

Regions of potential damage for airblast.   Airblast can do damage if the overpressure is high enough and there are two areas to be considered. They are referred to as Near Field and Far Field.

Near field.   This is the region in which the blast sound is transmitted directly from source to target. The potential for damage in the near field is small. Good planning, such as attention to the details of spacing, burden, stemming, explosive charge, delays, covering of detonating cork trunklines, and use of cord with minimal cord load minimizes damage potential.

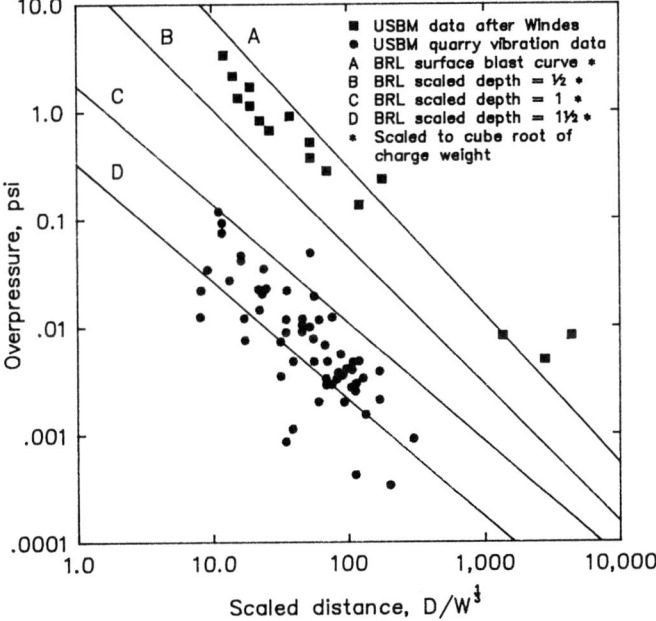

Figure 11.12   Combined data plot, overpressure versus scaled distance.

Far field and airblast focusing.   The far field as the name implies is the region far from the blast site, i.e., 4 to 20 miles. Neither direct transmission of airblast nor ground vibration can account for the effects produced. It represents a concentration or focusing of energy in a narrow region. Only the sound energy can explain this. Sound waves traveling up into the atmosphere are refracted back to the earth, concentrating in a narrow region to produce an intense overpressure.

How does airblast focusing occur? It is due to an atmospheric inversion in which the air temperature is warm above and cold below. The stronger the inversion the more intense the focusing may be. An additional factor is the effect of wind which can significantly influence the focusing.

Atmospheric Inversion

Under normal conditions, temperature decreases with height in the atmosphere, cooling at the normal lapse rate of 3.5°F for each 1000 ft of height. If this is reversed so that the temperature increases with height instead of cooling the condition, it is termed an inversion. Inversions are not uncommon.

The velocity of sound in air is a function of the temperature, increasing when the air is warmer, decreasing when the air is colder. The change is approximately 1 ft/sec for a temperature change of 1°F. In the cooler air of the upper atmosphere (normal conditions) sound velocity decreases, causing the sound waves to curve upward into the atmosphere away from the earth. The sound is absorbed in the atmosphere. This effect is illustrated in Fig. 11.13.

When an atmospheric inversion occurs, the air temperature increases with height, so the velocity of sound also increases. Because of the increased velocity, the sound waves curve downward toward the ground and may return to the earth, at some distance from the point of origin. Figure 11.14 illustrates the inversion condition and the downward curving of the sound rays in the atmosphere. The upward or downward curving of the sound rays is governed by Snell's Law.

Under the appropriate atmospheric inversion conditions the sound may con-

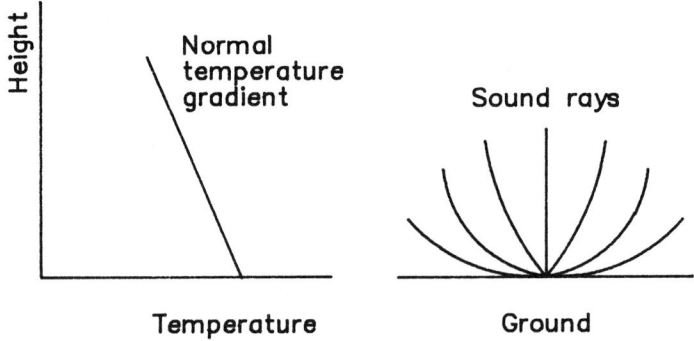

Figure 11.13   Normal atmospheric conditions.

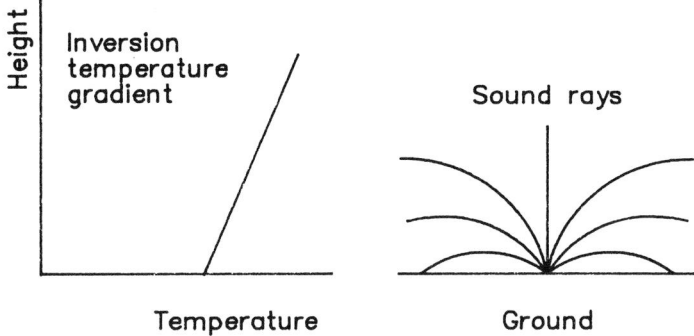

Figure 11.14   Atmospheric inversion.

centrate or focus in a narrow region and produce much higher sound levels than in adjacent regions on either side which may even have no sound. This effect is shown in Fig. 11.15.

Wind effect.   Wind can be a significant factor in airblast focusing. On the downwind side, the wind velocity will add to the velocity effect produced by the inversion and further increase the sound velocity. On the upwind side, the wind velocity will oppose the inversion velocity effect and decrease the sound velocity. If the wind is strong enough, the sound may be completely blown away from the upwind side. Figure 11.16 shows the wind effect.

Without the wind effect the focal region appeared as a distant ring surrounding the blast site. With the wind effect, the ring may be reduced to a crescent shaped focal region with much higher sound intensity. This is shown in Fig. 11.17.

Airblast focusing can result from the combination of an atmospheric temperature inversion and wind. The effect varies with height and must be evaluated at 1000 ft intervals of elevation, e.g., 1000 ft, 2000 ft, and so on. Thus, upper air mete-

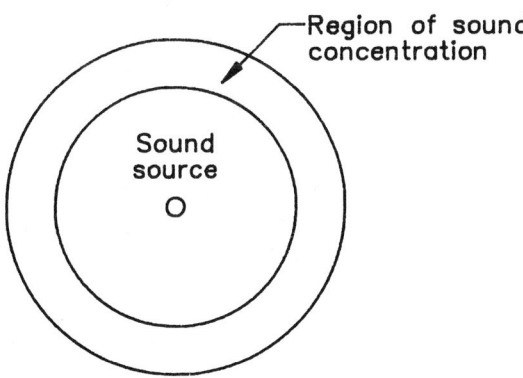

Figure 11.15   Airblast focusing.

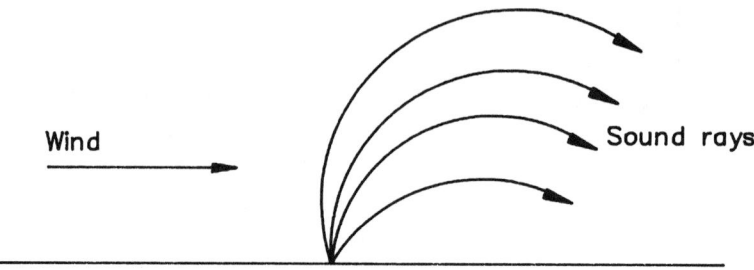

Figure 11.16   Wind effect.

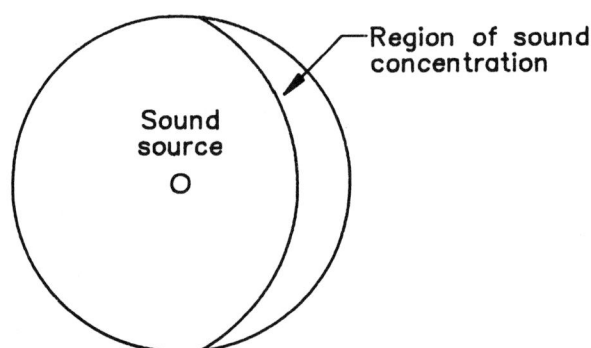

Figure 11.17   Airblast focusing plus wind effect.

orological data are necessary along with a sophisticated computer program to pro-cess it. This is not very practical for day to day operations. An example of an intense airblast focusing is shown in Fig. 11.18. Note that the concentration of sound in the range 60 to 65 kilofeet, approximately 12 to 13 miles. Sound may be amplified as much as 100 times normal.

PROCEDURES TO AVOID AIRBLAST FOCUSING

1. Don't shoot if there is an atmospheric inversion
2. Contact the local weather bureau to find out if there is an inversion
3. Radiation inversions commonly exist in the mornings, but normally disappear by noon. Hold the shot until the inversion has been dissipated. Frontal or air mass inversions tend to persist and do not go away.
4. Obtain wind information from the weather bureau. If the downwind direction is a populated sensitive area, avoid shooting; if it is unpopulated or industrial, shooting may be feasible.

Airblast focusing requires specialized atmospheric conditions and hence, is not a usual or common occurrence. It cannot occur at short distance and, in general, is not a problem for concern.

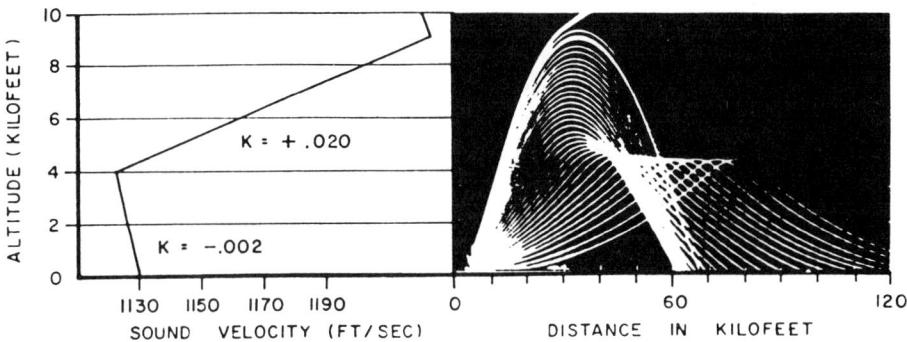

Figure 11.18 Sound focusing. (From Ballistics Research Laboratories, Report No. 124V.)

Preblast surveys.   Preblast inspections are being mandated more and more by various regulatory agencies, insurance companies and concerned operators.

Purpose.   A preblast inspection is intended to document the condition of a structure prior to its exposure to vibration from blasting. Cracks occur in practically all structures but the occupants are either unaware or only slightly aware of them.

The preblast survey will, therefore, educate the occupants with the fact that there are cracks. They are usually surprised by what the inspection reveals. It can also be used to verify or refute claims of damage resulting from vibration. So if a preblast inspection is called for, it must be done carefully and thoroughly.

Cracks in a structure are dynamic in nature. They change from season to season and are affected by factors such as temperature, humidity, wind, soil conditions, and overall structural integrity. Assuming reasonably stable soil conditions and structural integrity the diurnal temperature changes produce thermal stresses that may cause cracks to grow in length and width. Similarly the larger seasonal temperature changes from summer to winter and back produce significant thermal stresses. In addition, significant changes in humidity occur with seasonal change. The winter heating season normally causes a drying out of the structure resulting in shrinkage. In spring and summer, the humidity rises and the structure absorbs moisture and expands.

Environmental stresses cause cracks to occur in practically all structures. When blasting appears on the scene, and vibration shakes the house, the home owner begins to look things over and finds the prior existing environmentally produced cracks. The conclusion is automatic, the blasting caused the cracks, when in fact it did not.

Inspection procedure.   Begin with the name of the project, the principals involved, address, date, and time.

Describe the structure, 1 floor, 2 floors, frame painted, or masonry construction, porches, foundations. Gutters and downspouts and drainage should be noted. Unusual ground and soil conditions should be noted, i.e., lot has a 1 in 2 slope, slope failure, retaining wall inclined due to soil pressure, drainage problems, pooling of water.

The interior inspection is done room by room with each room designated and described, i.e., living room, plaster walls covered with wall paper, plastered and painted ceiling with plastered cove and wood molding at juncture of cove and wall paper.

Walls are numbered in a clockwise fashion 1, 2, 3, 4 (if more than 4 walls, state it and continue numbering) beginning with the wall to the left of the door by which you entered. Each wall is inspected, noting the number of windows, doors, and all cracks, holes, and so forth around the windows, doors, wall corners, and ceiling corners. Separation of dry wall joints or crazed plaster are to be noted, window moldings and floor and ceiling moldings should be checked for openings in mitered joints, straight joints and for separations from wall, floor or ceiling. Glass panes should be examined for cracks. A closet in a wall should be inspected after the wall inspection. Diagrams and sketches are helpful.

The floor is examined for openings between boards, loose, or squeaky boards, cracked boards, discoloration, and general wear of wood. If the floor is covered with tile examine tile for joint separations, cracked tile, holes, general wear. If carpeted state so and indicate the floor is not visible for inspection.

Ceiling is examined for cracked, loose or hanging plaster, water discoloration, and holes. Ceiling and floor are inspected with inspector's back to wall 1. This insures uniformity.

If a wall has no cracks, simply state wall 1—clear.

Fireplace in a wall should be examined for separation from the wall and for cracks in the masonry and plaster.

Attic inspection.   The attic is inspected as a room or series of rooms if so divided and finished. If unfinished, attention should be given to the roof rafters and keel board for separation. Also note cross bracing and knee walls and any open spaces where day light is visible.

Basement inspection.   When entering the basement by stairway from the first floor, the wall opposite the stairway is wall 1. Proceed with the inspection in clockwise fashion. Examine walls for cracks in mortar joints, cracks across block or stone, holes at pipe entry, cracks at juncture with next wall. Also note any evidence of settling. Note the plates on top of the wall, are they uniformity level, and so on. Note water stains and seepage.

In the floor inspection note location of floor drain, any cracked or broken areas or holes in the floor, evidence of water seepage. Also note if floor consists of sections.

Check ceiling for cross bracing between floor joists, cracked or broken joist, or rotted areas.

Check water pipes, heat pipes, and electrical conduit for any unusual circumstances.

Stairway inspection.   Stairway is considered as a unit going up. Describe right wall, left wall, ceiling. Stairs consist of tread and riser which are examined for cracks, marred finish, and so on.

Exterior inspection.   Inspection usually begins at front of house and proceeds in a clockwise direction inspecting each side in turn. State condition of masonry with regard to cracks, holes, poor mortar, shrinkage cracks, and so on. For frame construction, describe condition of siding, presence of cracks, openings at juncture of siding boards, warping, paint condition, new, old, flaking, and so on.

State number of windows and doors and examine frames and casing for cracks, warping, openings.

Fireplace, or furnace chimney, check for separation from wall, for cracks in masonry and joints and so on.

Porches, check for level and settling. Check for separation at juncture with wall. Check for cracked, separated and rotting boards.

Foundation, check for cracks, holes, settling, condition of mortar.

Gutter and downspouts, check for general condition and drainage. Does water drain onto roof, porch or side of house.

Garage.   Garage should be inspected as a room.

Walks and driveways.   All concrete walks, driveways, and so forth should be inspected for cracks, level conditions, holes, etc.

Unusual conditions.   Photographs should be made of any unusual conditions that are more severe than ordinary. For example, suppose a fireplace chimney settled because of an inadequate foundation and caused cracks in the basement masonry, the living room wall plaster, or both.

SAMPLE PREBLAST INSPECTION FORM

Project _____   Inspector _____

_____   Date _____

Building Inspected _____
                    Occupant or Building name

                    _____
                    Street Address

                    _____
                    City and State

Building Description

Number of floors _____

Construction (frame, brick, etc.) _____

Foundation (stone, block, etc.) _____

Porches _____

Roof _____

Gutters _____

Lot Location (level, sloping, etc.) _____

Other unusual factors _____

_____

_____

Room Inspection

General description L.R., walls painted, cove ceilings, etc. _____

_____

_____

Wall 1 (N-wall) _____

_____

_____

Wall 2 _____

_____

_____

Wall 3 _____

_____

_____

Wall 4 _____

_____

_____

Sketch of typical wall crack pattern.

Preblast survey reports.   Thoroughness and care are important in building inspection. Common sense based on knowledge of what to look for and where to look for it will insure an adequate inspection. The preblast survey report should be an adequate and complete description of what the inspection was able to document. It should be clearly written so that an independent examiner can readily understand what is being reported.

Effects of Blasting on Water Wells and Aquifers

Aquifers.   Whenever an aquifer or a water well exhibits change in a region where blasting occurs, the blasting may be cited as the cause. Under normal blasting circumstances it is highly improbable that the blasting will cause damage.

An aquifer is a rock formation with sufficient porosity and permeability to allow for the storage of water and the flow of water through the system. Charging of the aquifer results from rainfall percolating into the porous rock beneath the surface. Hence, the aquifer is intimately affected by the amount of rainfall and the seasonal conditions. The top of the water surface is known as the water table.

A well is a man made opening or hole drilled from the surface down into the aquifer to some depth below the water table. The level of the water in the well is the same as the level of the water table. During dry spells the water table may be lowered and wells with only a shallow penetration into aquifer may go dry. When the aquifer is recharged the well will return.

Vibration effects.   Vibration has frequently been blamed for problems that occur in wells but, recent investigations by the U.S. Bureau of Mines (P.R. Berger and Associates, 1982) indicate that blasting has little or no effect and that vibration below 2.0 in. per sec will not cause damage to a well.

Fracturing around a blast hole is limited to a radius of 20–40 blasthole diameters. For a 6-in. hole this is 10–20 ft and for a large blast hole 18 in. this is 30–60 ft.

In the Bureau of Mines investigations (P.R. Berger and Associates, 1982) 25 wells were drilled at four sites. They were tested both before blasting occurred and after blasting. When the blasting approached within 300 ft of a well at three of the sites the static water level (water table) dropped abruptly but was followed shortly by a significant improvement in well performance. At the fourth site, no change occurred. The time of the water level drop indicates that it was not a direct result of the blasting.

Particle velocities in the series of tests ranged from 5.4 ips to 0.84 ips resultant particle velocity.

The interpretation of the effects is that the storage capacity of the aquifer may have been increased thus enabling the aquifer to hold a larger volume of water. This resulted in a drop in the water level which soon recovered and increased the well performance.

The principal effect of blasting on water wells that are close by is that temporary turbidity may occur in the well. This turbidity condition passes quickly and is a temporary annoyance rather than a damage problem. Vibration levels below 2.0 in. per sec are not sufficient to cause damage to water wells.

Open cut.   If the purpose of the blasting is to excavate an open-cut then nearby wells might be affected. Several factors that would have to be considered are the depth of the open-cut and the direction of the underground flow from the aquifer into the well relative to the open-cut.

Under the right conditions the open-cut may have an undesirable effect on wells that are close by ranging from reduced capacity to total loss of the well. This, however, is not a blasting effect.

Assessment of Rippability versus Blasting

Definition.   Rippability refers to the ability of a bulldozer with a ripper hook to rip or tear out the rock. If the rock is relatively thin layered so that the ripper can cut into the bedding planes there is a good chance that it can be ripped. Massive rock that is thick layered is difficult to rip. Strength of the rock is a very important factor also. Strength of a rock is measured by its elastic moduli. Seismic velocity is also determined by the elastic moduli; hence, it is logical that rippability and seismic velocity be related.

Rippability and Seismic Velocity

Caterpillar Tractor has published tables of rippability based on seismic velocity and rock type. Seismic velocity or propagation velocity is the speed at which seismic waves travel through the rock. It varies from 1000 to 20,000 ft per sec with a lot of overlap for various rock types. For example, there are shales, sandstones, and limestones which have velocities of 9200 ft per sec. At this velocity the sandstone and the shale are marginally rippable with a D9N ripper but the limestone is nonrippable. It is necessary, therefore, to know the rock type as well as the seismic velocity.

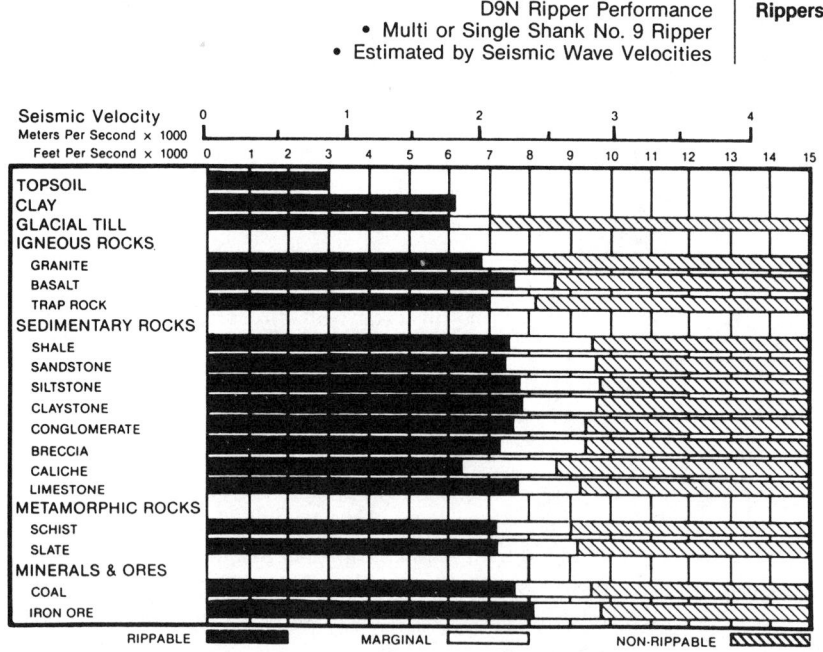

Figure 11.19   D9N ripper performance. (Courtesy of Caterpillar Tractor Company)

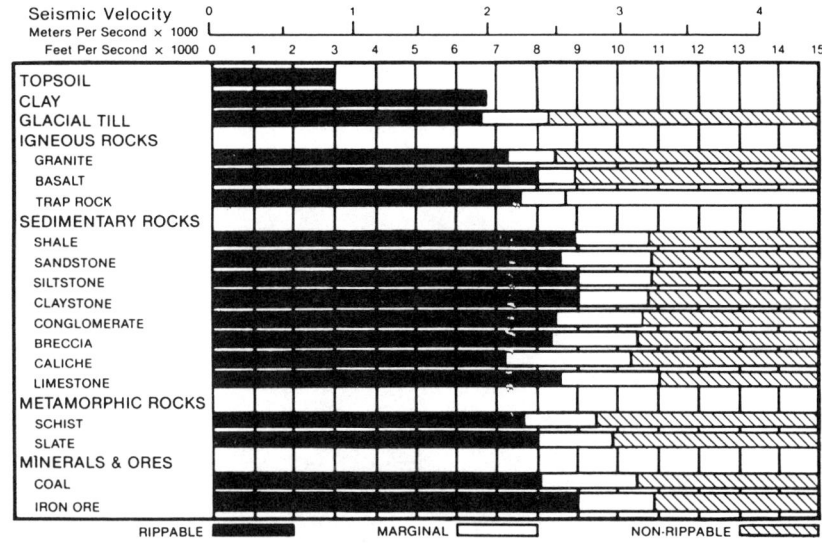

Figure 11.20 D10N ripper performance. (Courtesy of Caterpillar Tractor Company)

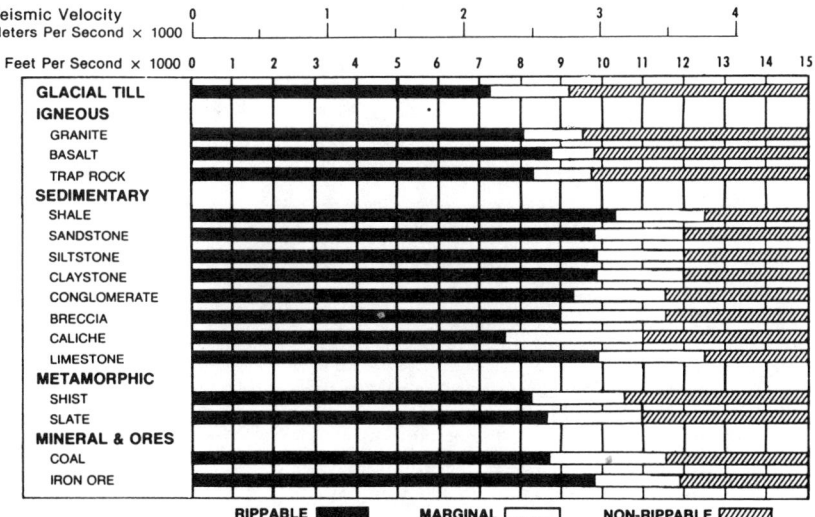

Figure 11.21 D11N ripper performance. (Courtesy of Caterpillar Tractor Company)

The rock type can be determined by test drilling the area. The seismic velocities can be determined by a shallow refraction seismic survey.

Rippability charts.   Rippability charts have been prepared by Caterpillar Tractor for various size bulldozers. These charts for D–9, D–10, and D–11 bulldozers are shown in Figs. 11.19 to 11.21.

PROBLEMS

1. Plot the following data on the log-log graph paper and draw in the envelope line.

Ds	PV
11	2.1
23	0.9
35	0.40
44	0.22
56	0.17

    a. What is the scaled distance Ds corresponding to a particle velocity of 1.0 ips?

2. The U.S. Bureau of Mines proposed safe vibration levels that used different velocities over two different frequency ranges. 2.0 ips above a certain frequency and 0.75 below that frequency. The value of that frequency is _____.

3. The U.S. Bureau of Mines Alternative Blasting Criteria used a displacement of 0.030 in. up to a frequency of 4 Hz and then at 4 Hz begins using a particle velocity of 0.75 ips. Show that these are equivalent at 4 Hz.

4. The Office of Surface Mining prescribes a particle velocity of 1.25 ips. maximum up to 300 ft while it only allows 0.75 ips at 5000 ft and beyond. Why?

5. Consider coal mining blasting, quarry blasting, and construction blasting.
    a. Which types of operation produce the lowest frequencies and what is the frequency range?
    b. Which produces the highest frequencies and what is the frequency range?
    c. Which is likely to produce the largest vibration in a house?
    d. Why?

6. What is the purpose of spectral analysis?

7. What is the purpose of response-spectra analysis?

8. Various fatigue type vibration tests show that in blast vibration fatigue is a

    1. major problem
    2. minor problem
    3. ordinary problem

9. Provide short answers for the following:

    a. The X-crack is also called a _____ crack.

    b. The X-crack is more likely to result from blasting that employs

            _____ charges
            _____ holes
            _____ distance

10. We are going to construct a synthetic seismogram.

    a. We begin with a single shot that is called a _____.

    b. This single blast data, in digitized form, is fed into a computer. A time interval starting at _____ and going up to _____.

    c. By this procedure, the ground motion is added to existing ground motion again and again producing a synthetic seismogram. The purpose of this is to _____.

11. What is meant by deck loading two charges in a hole? Draw a sketch.

12. When shooting near green concrete, what restrictions apply and what is the critical time period?

13. A simple rule of thumb for human response to vibration is _____.

14. Define the unit of sound level pressure, the decibel (equation).

15. A sound level of 60 dB is doubled.

    a. What is the new sound level in dB?

    b. What is the level of conversational speech?

    c. What is the limit in dB suggested by the U.S. Bureau of Mines to prevent plaster cracks?

16. Blast vibration is scaled by the square root of the charge. How is the sound scaled?

17. Airblast has two parts, one called noise and the other called _____.

18. The expected form of airblast damage that is first to occur is _____.

19. Ordinary borehole blasting is _____ to cause damage.

    a. likely

    b. unlikely

20. Airblast focusing in the far field is due primarily to _____ and a contributing factor is _____.

21. Provide short answers for the following questions:

    a. What is the general range of distance at which air blast focusing occurs? _____

    b. The overpressure may be amplified over normal as much as _____.

    c. Airblast damage is keyed to _____.

22. A road construction project is blasting 600 ft from a structure using a maximum charge per delay of 400 lb which produces a peak particle velocity of 0.82 ips. What particle velocity would you expect when the construction comes within 200 ft of structures?

REFERENCES

Ballistic Research Laboratories, "Forecasting the Focus of Airblasts Due to Meterological Conditions in the Lower Atmosphere," Report 1118, 1960.

Ballistic Research Laboratories, "Handbook for Predictions of Airblast Focusing," Report 1240, 1964.

BERGER, P.R. and ASSOCIATES, "Survey of Blasting Effects on Ground Water Supplies in Appalachia," vol. 1 and 2, Washington, DC, U.S. Bureau of Mines, 1982.

CRANDELL, F.J., "Ground Vibration Due to Blasting and Its Effect Upon Structures," *Journal of the Boston Society of Civil Engineers,* 49(2):152–168, 1949.

EDWARDS, A.T. and NORTHWOOD, T.D., "Experimental Blasting Studies on Structures," Ottawa, Canada, National Research Council, 1959.

GOLDMAN, D.E., "A Review of Subjective Responses to Vibrating Motion of the Human Body in the Frequency Range, 1 to 70 cycles per second," Naval Medical Research Institute Report No. 1, Project NM 004001, March 19, 1948.

"Handbook of Ripping," Caterpillar Tractor, Co., Peoria, IL.

KOERNER, R.M., DEISHR, J.N., and SCHELLER, R.P., "Laboratory Study of Cracking in Model Block Masonry Walls," *Proceedings of the Third Conference on Explosives and Blasting Technique,* Pittsburgh, PA: Society of Explosives Engineers, pp. 227–237, February 23, 1977.

LANGEFORS, U., WESTERBERG, H., and KIHLSTROM, B., "Ground Vibrations in Blasting," Parts I–III, *Water Power,* September 1958, 335–338; October 1958; 390–395; November 1958, 421–424.

LEET, L.D., "Effects Produced by Blasting Rock," Wilmington, DE: Hercules Powder Co., 1971.

MEDEARIS, K., "Dynamic Characteristics of Ground Motion Due to Blasting," *Bulletin of the Seismological Society of America,* 63(2):627–639, 1979.

NICHOLLS, H.R., JOHNSON, C.F., and DUVALL, W.I., "Blasting Vibrations and Their Effects on Structures," Bulletin No. 656, Washington, DC: U.S. Bureau of Mines, 1971.

ORIARD, L.L., "Observations on the Performance of Concrete at High Level Stresses from Blasting," *Proceedings of the Sixth Conference on Explosives and Blasting Technique,* 1980.

PERKINS, B., Jr. and JACKSON, W.F., "Handbook for Predictions of Airblast Focusing," Report No. 1240, Ballistic Research Labs, 1964.

PERKINS, B.J., Jr., LORRAIN, P.H., and TOWNSEND, W.H., "Forecasting the Focus of Airblasts Due to Meteorological Conditions in the Lower Atmosphere," Report No. 1118, Ballistic Research Labs, October 1960.

REIHER, H. and MEISTER, F.J., "Die Empfindlichkeit des Menschen gegen Erschuelterungen," (Sensitivity of Human Beings to Vibration) Forschung auf dem Gebert des Ingenieurwesens, Berlin: 2:381–386, February 1931.

SISKIND, D.E. and SUMMERS, C.R., "Blast Noise Standards and Instrumentation," Report No. TPR 78, Washington, DC: U.S. Bureau of Mines, 1974.

SISKIND, D.E., STAGG, M.S., KOPP, J.W., and DOWDING, C.H., "Structure Response and Damage Produced by Ground-Vibration from Surface Mine Blasting," Report No. RI 8507, Washington, DC: U.S. Bureau of Mines, 1980.

Sperry Univac, "Vibration Specifications for Computers," Minneapolis, MN, 1982.

STAGG, M.S., SISKIND, D.E., STEVENS, M.G., DOWDING, C.H., "Effects of Repeated Blasting on Wood-frame House," Report No. RI8896, Washington, DC: U.S. Bureau of Mines, 1984.

The Small Home, "Forty Reasons Why Walls and Ceilings Crack," Architects Small House Service Bureau of the United States, Inc., Minneapolis, MN, 4(8), 1925.

THOENEN, J.R. and WINDES, S.L., "Seismic Effects on Quarry Blasting," Bulletin No. 442, Washington, DC: U.S. Bureau of Mines, 1942.

WALTER, E.J., "Lithologic Variations and Vibration Effects," *Proceedings of the Seventh Conference on Explosives and Blasting Technique,* pp. 52–61, 1981.

WALTER, E.J., "Natural Variation of Vibration Level Associated with Blasting," *Proceedings of the Sixth Conference on Explosives and Blasting Technique,* pp. 331–340, 1980.

WISS, J.F. and PARMELEE, R.A., "Human Perception of Transient Vibrations," *Journal of the Structural Division, ASCE,* 100(ST4):773–787, 1974.

Appendix

Glossary of Blasting Terms

A-scale A sound level measurement scale. It discriminates against low frequencies. It approximates the human ear.

Acceleration A measure of force (F = ma). It is the time rate of change of velocity. It is measured in g's the acceleration of gravity.

Accessories General term applied to small equipment and tools used in conjunction with explosives, i.e., blasting machine, punches, wire, fuses, Galvanometer, etc.

Acoustic trace The line on the vibration record that records the sound level.

Acoustical impedance The mathematical expression for characterizing a material as to its energy transfer properties (the product of its unit density and its sound velocity (pV).

Acultural Vibration Vibration that is strange and unfamiliar to the observer.

Adit A nearly horizontal passage from the surface by which an underground mine is entered, as opposed to a tunnel.

Airblast A sound pressure wave from a blast traveling through the atmosphere.

Airblast focusing The concentration of sound energy in a small region at ground level due to refraction of the sound waves back to the earth from the atmosphere.

Air cushion A blasting technique wherein a charge is suspended in a borehole, and the hole tightly stemmed so as to allow a time lapse between detonation and ultimate failure of the rock. (No coupling realized.)

American Table of Distances Table showing distances that explosives must be stored from other explosives, inhabited buildings, railroads, highways, and magazines, according to amount of explosives stored. Usually called only Table of Distances.

Amplitude The height of a vibration or wave above the zero line on a vibration record. It usually refers to the maximum value.

ANFO Ammonium Nitrate–Fuel Oil Mixture. Used as a blasting agent.

AN prills Small spheres or pellets of ammonium nitrate, as opposed to flaked, granular, or powdered ammonium nitrate.

Back The roof or top of an underground opening. Also, used to specify the ore between a level and the surface, or that between two levels.

Back break Rock broken beyond the limits of the last row of holes.

Bedding planes Rock formation formed by layering of rock as it was deposited, as in igneous flows or in separated sedimentary deposits.

Bench The horizontal ledge in a quarry face along which holes are drilled vertically. Benching is the process of excavating whereby terraces or ledges are worked in a stepped shape.

Blast The operation of breaking rock by means of explosives. Shot is also used to mean blast.

Blasting agent Any material or mixture, consisting of a fuel and oxidizer intended for blasting, not otherwise classified as an explosive and in which none of the ingredients are classified as an explosive, provided that the finished product, as mixed and packaged for use or shipment, cannot be detonated by means of a No. 8 test blasting cap when confined.

Blasthole (Borehole) A hole drilled in rock or other material for the placement of explosives.

Blockhole A hole drilled into a boulder to allow the placement of a small charge to break the boulder.

Body waves Seismic wave that travels through the mass or body of a rock material.

Booster A chemical compound used for intensifying an explosive reaction. A booster does *not* contain an initiating device.

Boot-leg A situation in which the blast fails to cause total failure of the rock because of insufficient explosives for the amount of burden, or caused by incomplete detonation of the explosives. That portion of a borehole that remains relatively intact after having been charged with explosive and fired.

Bridging Where the continuity of a column of explosives in a borehole is broken, either by improper placement, as in the case of slurries or poured blasting agents, or where some foreign matter has plugged the hole.

Buffer Previously shot material, not removed, lying against a face to be shot.

Bulk strength Refers to the strength of a cartridge of dynamite in relation to the same sized cartridge of straight Nitroglycerine dynamite.

Burden Generally considered the distance from an explosive charge to the nearest free or open face at the time the hole detonates. Technically, there may be an apparent burden and a true burden, the latter being measured always in the direction in which displacement of broken rock will occur following firing of an explosive charge.

C-scale A sound level measurement scale that has only slight discrimination at the low frequencies.

Centers The distance measured between two or more adjacent blastholes without reference to hole locations as to row. The term has no association with the blasthole burdens.

Charge weight The amount of explosive charge in pounds.

Chambering More commonly termed springing. The process of enlarging a portion of a blasthole (usually the bottom) by firing a series of small explosive charges.

Collar The mouth or opening of a borehole, drill steel, or shaft. Also, to collar in drilling means the act of starting a borehole.

Compressional wave A seismic wave whose motion is compression-dilatation or push-pull, generated by rock's resistance to compression.

Condensor-discharge A blasting machine which uses batteries to energize a series of condensors, whose stored energy is released into a blasting circuit.

Connecting wire Any wire used in a blasting circuit to extend the length of a leg wire or leading wire.

Connector Refers to a device used to initiate a delay in a Primacord circuit, connecting one hole in the circuit with another, or one row of holes to other rows of holes.

Coupling The act of connecting or joining two or more distinct parts. In blasting the reference concerns the transfer of energy from an explosive reaction into the surrounding rock and is considered perfect when there are no losses due to absorption or cushioning.

Coyote blasting The practice of drilling blastholes (tunnels) horizontally into a rock face at the foot of the shot. Used where it is impractical to drill vertically.

Crest The top of the face created by a previous shot. The maximum amplitude of a wave in the upward direction above the zero line.

Cultural vibration Vibration that is commonplace and familiar to the observer.

Cushion blasting The technique of firing of a single row of holes along a neat excavation line to shear the web between the closely drilled holes. Fired after production shooting has been accomplished.

Cut More strictly it is that portion of an excavation with more or less specific depth and width, and continued in similar manner along or through the extreme limits of the excavation. A series of cuts are taken before complete removal of the excavated material is accomplished. The specific dimensions of any cut is closely related to the material's properties and required production levels.

Cut off Where a portion of a column of explosives has failed to detonate because of bridging, or to a shifting of the rock formation due to an improper delay system.

Decibel (dB) The unit of sound level measurement.

Deck In blasting a smaller charge or portion of a blasthole loaded with explosives that is separated from the main charge by stemming or air cushion.

Deflagration An explosive reaction that consists of a burning action at a high rate of speed along which occur gaseous formation and pressure expansions.

Delay The term used to describe a blasting cap which does not fire instantaneously but has a predetermined built-in lag or delay.

Delay blasting Blasting that uses delays or delay caps.

Delay element That portion of a blasting cap which causes a delay between the instant of impressment of electrical energy on the cap and the time of detonation of the base charge of the cap.

Density The mass of an explosive per unit volume. Expressed in gm/cc. Water has a density of 1.0 gram per cubic centimeter.

Detonating cord A plastic covered core of high velocity explosives used to detonate charges of explosives in boreholes and under water, e.g., Primacord.

Detonation An explosive reaction that consists of the propagation of a shock wave through the explosive accompanied by a chemical reaction that furnishes energy to sustain the shock-wave propagation in a stable manner, with gaseous formation and pressure expansion following shortly thereafter.

Dip The angle at which strata, beds, or veins are inclined from the horizontal.

Displacement The amount of motion associated with vibration of waves, measured in inches.

Double priming A blasthole containing two priming units, usually on the same time delay. They are usually placed one near the top and one near the bottom of the blasthole.

Downlines Primacord lines running from the top of the holes. The primer is attached to the bottom end and additional primers may be slid down the cord in the case of decking and/or multiple priming.

Drop ball Known also as a Headache Ball. An iron or steel weight held on a wire rope that is dropped from a height onto large boulders for the purpose of breaking them into smaller fragments.

EBC (Electric Blasting Cap): May be instantaneous or delayed. Used to initiate primers or detonating cord.

Elastic limit The limit of elasticity or strength of a rock. Stress below the elastic limit generates elastic waves (seismic waves) while stress above the elastic limit produces rock breakage.

Elasticity The property of material that enables it to regain its original size and shape after it has been deformed.

Energy ratio A standard for damage caused by vibration from blasting. Also written as ER and defined as (acceleration in ft/sec/frequency).

Explosion A thermochemical process whereby mixtures of gases, solids, or liquids react with the almost instantaneous formation of gaseous pressures and near sudden heat release. There must always be a source of ignition and the proper temperature limit reached to initiate the reaction. Technically, a boiler can rupture but cannot explode.

Explosive Any chemical mixture that reacts at high speed to liberate gas and heat and thus cause tremendous pressures. The distinctions between High and Low Explosives are twofold; the former are designed to detonate and contain at least one high explosive ingredient; the latter always deflagrate and contain no ingredients which by themselves can be exploded. Both High and Low Explosives can be initiated by a single No. 8 blasting cap as opposed to Blasting Agents which cannot be so initiated.

Explosive charge The quantity of explosive that is to be detonated.

Explosive decks Explosives placed in certain areas of the hole separated by drill cuttings.

Face The end of an excavation toward which work is progressing or that which was last done. It is also any rock surface exposed to air.

Far field The region sufficiently far from a sound source in which direct transmission of sound waves is negligible.

Fatigue The weakening or failure of a material because of repeated vibration or strain.

Fire In blasting it is the act of initiating an explosive reaction.

Floor The bottom horizontal, or nearly so, part of an excavation upon which haulage or walking is done.

Flyrock Rock that is propelled into the air by the force of the explosion. Usually comes from prebroken material on the surface or upper open face. Flyrock is an indicator of wasted energy.

Fracture Literally, the breaking of rock without movement of the broken pieces.

Fragmentation The extent to which rock is broken into small pieces by primary blasting.

Frequency The number of vibrations or complete oscillations occurring in one second designated Hertz, (Hz) or cycles per second (cps).

Fuel In explosive calculations it is the chemical compound used for purposes of combining with oxygen to form gaseous products and cause a release of heat.

Galvanic action Currents caused when dissimilar metals contact each other or through a conductive medium. This action may create sufficient voltage to cause premature firing of an electric blasting circuit, particularly in the presence of salt water.

Galvanometer A device containing a silver chloride cell which is used to measure resistance in an electric blasting circuit.

Grade In excavation, it specifies the elevation of a roadbed, rail, foundation, and so on. When given a value such as percent or degree grade it is in the amount of fall or inclination compared to a unit horizontal distance for a ditch, road, etc. To Grade means to level ground irregularities to a prescribed level.

Gram atom The unit used in chemistry to express the atomic weight of an element in terms of grams (weight).

Ground calibration Determination of the vibration transmission characteristics of a region.

Hardpan Boulder clay, or layers of gravel found usually a few feet below the surface and so cemented together that it must be blasted or ripped in order to excavate.

Hertz The frequency of vibration written Hz or cycles per second (cps).

Highwall The bench, bluff or ledge on the edge of a surface excavation and most usually used only in coal strip mining.

Human response The reaction of a person to different vibration levels.

Initiation The act of detonating a high explosive by means of a mechanical device or other means.

Initiator A device or product used to transmit and/or supply heat and/or shock to start an explosion.

Inversion An abnormal atmospheric·condition such that the air temperature increases with height instead of decreasing.

Joints Planes within rock masses along which there is no resistance to separation and along which there has been no relative movement of the material on each side of the break. They occur in sets, the planes of which are generally mutually perpendicular. Joints, like stratification, are often called partings.

Jumbo A machine designed to contain two or more mounted drilling units which may or may not be operated independently.

Kelly bar A two-piece drill steel. The inner steel may be withdrawn, allowing the loading of a cardboard casing and/or explosives, while the outer steel prevents rocks or cuttings bridging the blasthole. Used mainly in Florida.

Lead wire The wires connecting the electrodes of an electric blasting machine with the final leg wires of a blasting circuit.

LEDC *Low Energy Detonating Cord.* Used to initiate nonelectric caps at the bottom of boreholes.

Linear scale A sound level measurement scale that is nonweighted so that there is little or no discrimination at low frequencies.

Longitudinal component That component of vibration which produces motion in the direction of a line joining the vibration source and the seismograph.

Longitudinal trace The line on the vibration record that records the longitudinal component of motion.

Low order Used to describe a condition of detonation that is not as rapid or complete as it should be.

Mat Used to cover a shot to hold down flying material; usually made of woven wire cable or rope.

Millisecond delay caps Delay electric caps which have a built-in delay element, usually 25/1000th of a second apart, consecutively. This timing may vary from manufacturer to manufacturer.

Misfire A charge, or part of a charge, which for any reason has failed to fire as planned. All misfires are to be considered extremely dangerous until the cause of the misfire has been determined.

Mole A unit in chemical technology equal to the molecular weight of a substance expressed in grams (weight).

Muck pile The pile of broken material or dirt in excavating that is to be loaded for removal.

Mud cap Referred to also as Adobe or Plaster Shot. A charge of explosive fired in contact with the surface of a rock after being covered with a quantity of mud, wet earth, or similar substance, no borehole being used.

MS connector A nonelectric millisecond delay device used with detonating cord for delaying shots from the surface.

Near field The region near the sound source in which there is direct transmission of sound.

Normal distribution The bell shaped symmetrical curve used in statistical analysis to generalize the relative frequency of occurrence of events.

Normal lapse rate Rate of decrease of temperature upward through the atmosphere. The average value is 3.5° F/1000 ft.

Open pit A surface operation for the mining of metallic ores, coal, clay, and so on.

Overbreak Excessive breakage of rock beyond the desired excavation limit.

Overburden The material lying on top of the rock to be shot; usually refers to dirt and gravel, but can mean another type of rock; e.g., shale over limestone.

Overpressure The pressure generated by a sound wave which produces variation in the atmospheric pressure. Overpressure is measured in psi or decibels.

Over shot Condition resulting from more than the necessary amount of explosives. Usually characterized by excesses of fragmentation, flyrock, and noise.

Oxidizer A supplier of oxygen.

Particle velocity The velocity at which the earth vibrates, measured in inches per second.

Peak particle velocity The maximum particle velocity.

Period The time for one complete vibration or oscillation of a wave measured in seconds.

Permissible Explosives having been approved by the U.S. Bureau of Mines for nontoxic fumes, and allowed in underground work.

Powder Any of various solid explosives.

Premature A charge which detonates before it is intended to.

Presplitting Stress relief involving a single row of holes, drilled along a neat excavation line, where detonation of explosives in the hole causes shearing of the web of rock between the holes. Presplit holes are fired in advance of the production holes.

Primary blast The main blast executed to sustain production.

Primer An explosive unit containing a suitable firing device that is used for the initiation of an entire explosive charge.

Propagation The movement of a detonation wave, either in a column or from hole to hole.

Propagation velocity The velocity at which a vibration or seismic wave travels outward from the source. It is measured in thousands of feet per second.

Quarry An open or surface mine used for the extraction of rock such as limestone, slate, building stone, and so on.

Response-spectra A methodology in which the response of a structure to different frequencies can be estimated mathematically.

Rip-Rap Coarse sized rocks used for river bank, dam, and so forth, stabilization to reduce erosion by water flow.

Round A group or set of blastholes constituting a complete cut in underground headings, tunnels, etc.

Safe limit The amount of vibration that a structure can safely withstand. Vibration below this limit has a very low probability of causing damage. Vibration above this limit has a reasonable probability of causing damage.

Scaled distance Factor of distance and quantity of explosive which relates to seismic disturbance.

Seam A stratum or bed of mineral. Also, a stratification plane in a sedimentary rock deposit.

Secondary blasting Using explosives to break up larger masses of rock resulting from the primary blasts, the rocks of which are generally too large for easy handling.

Seismic velocity The same as propagation velocity. The velocity at which a seismic wave travels outward from its source.

Seismic waves Waves that travel through the earth.

Seismograph An instrument that measures and supplies a permanent record of earthborne vibrations induced by earthquakes, blasting, and so forth.

Seismograph trace A line on a seismograph record showing the vibration of the seismic wave.

Sensitizer The ingredient used in explosive compounds to promote greater ease in initiation or propagation of the reactions.

Sensor A device that senses or measures the vibration of the ground.

Shear wave A seismic wave whose motion is at right angles to the direction of travel. It is generated by the rock's resistance to shear or change in shape.

Shot firer Also referred to as the Shooter or Blaster. The person who actually fires a blast. A Powderman, on the other hand, may charge or load blastholes with explosives but may not fire the blast.

Shunt A piece of metal connecting two ends of leg wires to prevent stray currents from causing accidental detonation of the cap. The act of deliberately shorting any portion of an electrical blasting circuit.

Sinking-cut A round drilled, loaded, and timed to be lifted vertically, due to the fact that no open face is available.

Slope Used to define the ratio of the vertical rise or height to horizontal distances in describing the angle a bank or bench face makes with the horizontal. For example, a 1 1/2 to 1 slope means there would be a 1 1/2 ft rise to each 1 ft or horizontal distance.

Snake hole A hole drilled or bored under a rock or tree stump for the placement of explosives.

Sound level The value of the sound level pressure in psi or in decibels.

Spacing In blasting, the distance between boreholes or charges in a row.

Spectral Analysis A method of analyzing the vibration frequencies present in a vibration record. It is basically a Fourier analysis.

Steady state velocity The chemically compounded rate of detonation of an explosive. Governed by diameter, degree of confinement, temperature, and so forth.

Stemming The inert material, such as drill cuttings, used in the collar portion (or elsewhere) of a blasthole so as to confine the gaseous products formed on explosion. Also, the length of blasthole left uncharged.

Strength Refers to the energy content of an explosive in relation to an equal amount of ANFO.

Stratification Planes within sedimentary rock deposits formed by interruptions in the deposition of sediments.

Strike The course or bearing of the outcrop of an inclined bed or geologic structure on a level surface.

Subdrill To drill blastholes beyond the planned grade lines or below floor level.

Surface waves Seismic waves that travel over the surface of the earth or rock rather than traveling into the rock mass.

Swell factor The ratio of the volume of material in its solid state to that when broken.

Tamping The process of compressing the stemming or explosive in a blasthole.

Throw and heave This has to do with the displacement of rock as a result of a detonation and the resulting expansion of gases.

Toe The burden or distance between the bottom of a borehole to the vertical free face of a bench in an excavation.

Trace A line on a vibration record.

Trace amplitude The amplitude of a seismic wave on any of the traces of the vibration record.

Transducer A device which can change energy from one form to another, i.e., mechanical energy into electrical energy.

Transverse A direction of motion at right angles to another direction of motion.

Transverse trace The trace on a vibration record that records motion at right angles to the line joining the vibration source and the seismograph.

Transverse wave A seismic wave that vibrates at right angles to the line joining the vibration source and the seismograph.

Trough The maximum amplitude of a wave in the downward direction below the zero line.

Trunk—Trunkline Detonation cord line on the surface, to which the downline or Primadet line is tied prior to firing.

Under shot A condition resulting from not enough explosive or a pattern size too large for the amount of explosive used. Usually characterized by poor fragmentation and lack of movement.

Velocity The rate of change of distance with time. The rate of detonation.

Vertical The direction perpendicular to the earth's gravity field. In this case, the up and down motion of the earth's vibration.

Vertical trace A line on the vibration record that shows the up and down motion of the earth's vibration.

Vibration crack A tension crack characteristic of material failure because of vibration. It has X-shape.

Vibration parameters Those physical quantities that are used to describe the vibration. These parameters are displacement, velocity acceleration, and frequency.

Vibration problem The human problem that arises when vibration is large enough to be felt.

Wave length The distance between two successive crests or troughs of a wave.

Wave parameters Those mathematical quantities that are used to decribe wave motion. These parameters are amplitude, period, frequency, wave length, and so on.

X-crack A tension crack characteristic of material failure due to vibration.

Index